Springer
Berlin
Heidelberg
New York
Barcelona
Budapest
Hong Kong
London
Milan
Paris
Santa Clara
Singapore
Tokyo

Jamshed R. Tata

Hormonal Signaling and Postembryonic Development

 Springer

Jamshed R. Tata

National Institute for Medical Research
London, United Kingdom

ISBN: 3-540-64259-5 Springer-Verlag Berlin Heidelberg New York
Biotechnology Intelligence Unit
Library of Congress Cataloging-in-Publication data

Tata, Jamshed R. (Jamshed Rustom)
    Hormonal signaling and postembryonic development / Jamshed R. Tata.
        p. cm. — (Biotechnology intelligence unit)
    Includes bibliographical references and index.
    ISBN 1-57059-512-7 (alk. paper) √
    1. Endocrinology, Developmental. 2. Cellular signal transduction.
    3. Hormones — Physiological effect.   I. Title. II. Series.
    [DNLM: 1. Signal transduction — physiology. 2. Cell
Differentiation. 3. Metamophosis, Biological — physiology. 4. Gene
Expression Regulation — physiology. 5. Cell Communication-
-physiology. 6. Morphogenesis.   QH 601 T216h 1997]
QP187.6.T38 1997
573.4'4 — dc21
DNLM/DLC
for Library of Congress   97-31479
                                                                CIP

Typesetting: R.G. Landes Company Georgetown, TX, U.S.A.

SPIN 10673821   31/3111- 5 4 3 2 1 0 - Printed on acid-free paper

Even a cursory look at one of the biomedical databases will confirm the enormous number of reviews and monographs that are now rapidly accumulating on diverse aspects of cellular signaling and developmental biology. So, why another book on these topics? Two principal reasons suffice to justify my decision to write about hormonal signaling and postembryonic development. First, the vast majority of the research in the last 10-15 years in developmental biology has been directed towards early embryogenesis. Impressive as the recent advances are in defining the genetic, cellular and molecular mechanisms underlying early development, they do not throw much light on how the adult phenotype is acquired later during postembryonic development. A major distinction between the early and late developmental processes is the obligatory control of the latter by hormonal and other extracellular signals. This constitutes the second reason justifying this monograph. It is an attempt to consider postembryonic development in close juxtaposition with mechanisms of hormonal signaling.

The first half of the book discusses hormone action, receptors, signal transduction and hormonal regulation of transcription, based on material drawn from research undertaken in a large number of centers throughout the world. The second half of the book deals with postembryonic development as a function of hormonal signaling, which relies heavily (but not exclusively) on experimental studies from my own laboratory over the last 30 years. It is hoped that the younger researcher or newcomer in this field will find the brief historical account at the beginning of each of the eight chapters useful. Over ten thousand original research articles, reviews and conference publications were screened for this monograph. Only a fraction of these, which are highly significant or relevant, could be included in a book of this size. A personal bias in the selection of literature as cited is thus unavoidable. Nevertheless, I have tried as much as possible to include, without losing sight of the wider perspective, the facts and literature illustrating the important or general principles that link hormonal signaling to the regulation of postembryonic development. I trust that the many investigators whose work I admire will be indulgent if I have failed to do it justice in my citation of their publications.

Most importantly, I wish to acknowledge the contributions of my research associates, graduate students, postdoctoral fellows, staff and visiting scientists over the years, too numerous to be cited individually here, without which it would have been impossible to write this book. It would also have been impossible to prepare this book without the help of Ena Heather, to whom I am most grateful. I also wish to acknowledge the expertise of John Satchell and other members of the PhotoGraphics Division of this Institute in preparing the illustrations.

I would also like to thank Academic Press, Elsevier Science, Carolina Biological Supplies Co., U.S. National Academy of Sciences, Macmillan Magazines, The Endocrine Society, Cold Spring Harbor Laboratory Press and The Pontifical Council for permission to reproduce published material.

Jamshed R. Tata

INTRODUCTION

As life evolved from simple unicellular to multicellular organisms of increasing complexity, these became progressively more dependent on mechanisms of sensory perception and communication for survival and reproduction. Such mechanisms involved both the external environment and internal milieu which in turn, necessitated signaling molecules. In simple eukaryotes, these molecules could be abundant nutrients, ions, or waste products, such as amino acids, sugars, Ca^{2+} and phosphate, which would also serve as signals for communication between individual organisms. In increasingly complex multicellular eukaryotes intercellular communication would be required more and more for most physiological processes, so that it became essential to synthesize specific molecules, either in all cells or specialized cells. The primitive extracellular signals, such as indole acetic acid and serotonin which are found in plants and animals, had to be integrated into intracellular signaling networks generated to coordinate responses to the extracellular signals. These intracellular molecules would be derived from diverse cellular constituents, and which include cyclic nucleotides, phospholipids, phosphorylated proteins, and nitric oxide. An essential element in the successful integration of extra- and intracellular signaling mechanisms are receptors, which could either be cellular components that may have served other functions in primitive organisms or which may have evolved more specifically to respond to signals more efficiently.

The last decade has seen dramatic progress in defining the molecular mechanisms underlying early development following fertilization of the egg. Many of these involve the expression of genes encoding signaling molecules and their receptors which, in turn, are essential for cell-cell communication during early embryogenesis. Most signals for early embryogenesis are produced in all cell types (or not in specialized cells) and can be classified as autocrine or paracrine, i.e., those that act on the same cells that produce them or on neighboring cells. A good example is activin, which regulates mesoderm induction. Compared with the impressive recent advances in our understanding of early embryogenesis, we know relatively little about the molecular genetics of postembryonic development, a more complex process whose major function is the establishment of the adult phenotype. Yet, this later period of development is particularly dependent on extracellular signals for its initiation, maintenance and completion. A large majority of signaling molecules regulating postembryonic development are hormones, i.e., produced in specialized cells and transported in the blood to various, often remote, target cells. Interestingly, activin which acts as a paracrine signal during early development is a hormone in adult vertebrates with quite different function, namely the regulation of reproductive activity. It is one of many examples of the same signaling molecule being put to different uses during evolution and development.

Among the wide range of postembryonic developmental systems dependent on signaling mechanisms, perhaps none is more dramatic than metamorphosis. Many of the morphogenetic and functional changes that occur during metamorphosis have their counterparts, perhaps not as dramatically, in fetal development in mammals. A major drawback in clearly defining the roles of molecular mechanisms underlying the hormonal regulation of mammalian fetal and perinatal development is that it is not possible to separate spatially and temporally the signals that originate in the developing fetus from those in the mother. Hence, because of its occurrence in free-living embryos, metamorphosis in both vertebrates and invertebrates is an ideal model for studying hormonal regulation of postembryonic development. A characteristic feature of metamorphosis that distinguishes it from early embryogenesis is extensive cell death and selective tissue regression, thus also providing an ideal model for studying the hormonal regulation of programmed cell death or apoptosis during postembryonic development.

In the first chapter the reader will find a brief account of the similarities and differences between early embryogenesis and postembryonic development, followed by three chapters covering the major features of growth and developmental hormones, their receptors, some of their comparative and evolutionary aspects, and the hormonal regulation of transcription. Next is the question of the important morphological and biochemical changes of genetic reprogramming and its hormonal control during amphibian metamorphosis. Since this postembryonic developmental process is obligatorily initiated by thyroid hormone, a more detailed account will be given about its receptors, the genes encoding them and their function as members of the family of nuclear receptors in their capacity as ligand-activated transcription factors. Two chapters emphasize the importance of auto- and cross-regulation of nuclear receptor genes and its significance in the regulation of not only metamorphosis but postembryonic development in general. The last chapter is devoted to the hormonal regulation of programmed cell death, a major feature of postembryonic development. It is hoped that the reader will find in this book an attempt to bring together two intensively researched fields of modern biology, namely development and molecular signaling, with a more specific purpose of presenting a perspective for further understanding of hormonal control of postembryonic development.

CONTENTS

CHAPTER 1

Early Embryogenesis and Postembryonic Development

Extraordinary advances have been made in the last decade in our understanding of the molecular mechanisms underlying the early stages of embryonic development. These, in turn, have greatly influenced current thinking on such important processes as gene expression, morphogenesis, patterning and signal transduction. By comparison, considerably less is known about the regulation of gene expression during the relatively lengthy period of development after the completion of embryo formation that precedes the acquisition of adult structures and functions. The reader is referred to several excellent textbooks of developmental biology for detailed accounts of early and late development.[1-5]

Both Early Embryogenesis and Postembryonic Development Are Characterized by Extensive Gene Activation

More than in any other area of developmental biology, the exploitation of the technical and conceptual advances in molecular genetics in the last 10-15 years have had a most profound impact on our understanding of early embryogenesis. This can perhaps be best illustrated by a count of the number of papers published between 1977 and 1992 in major journals dealing with molecular genetics of early embryonic development of one organism alone, namely *Xenopus*. As depicted in Fig. 1.1, these went up 35 times, from 17 to nearly 600, during these 15 years. In the same period, publications on questions relating to gametogenesis, sex determination and postembryonic development in the same organism increased less than 3.5 times, from about 45 to 150. Nevertheless, one fact clearly emerges from work on all invertebrate and vertebrate species in that different stages of development are characterized by more or less discreet periods

Hormonal Signaling and Postembryonic Development,
by Jamshed R. Tata. © 1998 Springer-Verlag and R.G. Landes Company.

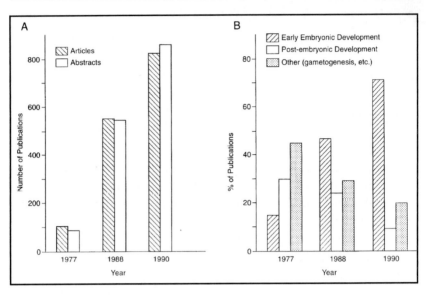

Fig. 1.1. A literature search of publications between 1977 and 1992 dealing with development of the frog *Xenopus laevis*.
A. The rapid rise in total number of publications.
B. The increase expressed separately as publications on early embryogenesis, postembryonic development, gametogenesis and other topics.

(varying according to species) of intense and selective activation of gene expression.

Figure 1.2 is an oversimplified representation of three important periods of relative gene activity from oogenesis to the adult stage. It is now well-known that vastly diverse mRNAs are stored in the egg and whose translation is activated upon fertilization.[2,3,6,7] In general, the eggs of oviparous organisms are considerably larger than those of mammals and that amount of total RNA stored is a function of the size of the egg of a given species. Thus, as shown in Table 1.1, the size of the *Xenopus* egg is almost 4,000 times that of the mouse or the human egg (the mass of egg per total body weight of *Xenopus* egg is almost 1,000,000 times that of the human egg!). The same holds true for the amount of total RNA per egg.[8] The larger size and mass of eggs of oviparous animals is also a reflection of the rapid growth and development of their free-living embryos. However, what is particularly striking is that the amount of RNA per unit mass or volume of the egg, irrespective of whether the organism is oviparous or viviparous, vertebrate or invertebrate, is remarkably constant (3-6 fg/μm).[3]

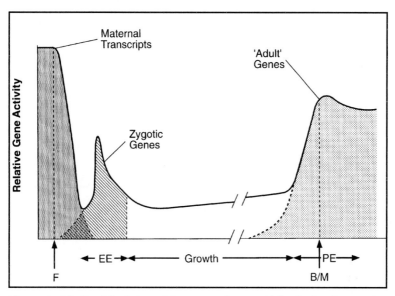

Fig. 1.2. An idealized depiction of the extent of gene activity during three major developmental periods. The first is immediately after fertilization and is largely due to maternally derived transcripts stored in the egg. The second is the activation of zygotic genes usually occurring after gastrulation; transcripts encode proteins necessary for cellular differentiation during early embryogenesis. The third represents the activation of genes during postembryonic, larval or fetal developmental periods, many of which encode specialized products (see Table 1.3) synthesized in a highly tissue-specific manner. The area under each peak is the product of a number of different genes and the abundance of their transcripts. Generally, early embryogenesis is characterized by many different genes with relatively low copy number transcripts, while during postembryonic development a few genes encode highly abundant mRNAs.

The bulk of RNA in the unfertilized egg is made up of ribosomal RNA and histone mRNAs. The pioneering studies of Brown and Gurdon established a firm biochemical basis for the changing pattern of RNA synthesis in the developing oocyte and early embryo of *Xenopus*.[9] Later work based on nucleic acid hybridization and characterization of mRNAs revealed another important feature of the composition of egg mRNA, namely that a small amount of messenger RNA comprises transcripts from 3-4,000 genes, irrespective of the organism.[2,7] Similar patterns of translation in vitro of mRNAs, and even protein phosphorylation, have been observed for eggs from very different organisms. Many of the transcripts present in high copy number are known to code for cytoskeletal proteins, chromosomal proteins, RNA and DNA polymerases, translational machinery

Table 1.1. Total content and concentration of RNA in eggs of some species

Organism	Volume of egg	RNA Content	
	(μm^3)	per egg (ng)	per vol. (fg/μm^3)
Rabbit	1.0×10^6	6.0	6.0
Mouse	1.8×10^5	0.5	2.8
Xenopus	7.2×10^8	4000.0	5.7
Sea urchin	7.0×10^5	4.3	6.1
Starfish	8.0×10^5	4.8	6.0

and cell membranes, whereas the less abundant transcripts code for such components as homeobox gene products, transcription factors, growth factors and receptors.[4,10-13] The application of techniques of in situ hybridization and immunocytochemistry to the analysis of the egg before and after fertilization has dramatically revealed how the unequal distribution of mRNAs in the egg determine positional information which is essential for generating the correct morpho-genetic patterning during early embryogenesis.[14-18] A number of pro-teins encoded by mRNAs stored in the egg have been shown to play a role in the onset of transcription of zygotic genes at the very early stages of embryogenesis as, for example, up to mid-blastula transi-tion or gastrulation in amphibia and birds.[6,19-21] This fact provides a molecular explanation for the phenomenon known to developmen-tal biologists for several decades of "maternal effects" in early em-bryogenesis.

Some of the maternally derived transcripts also encode tran-scription factors and other regulatory proteins that activate zygoti-cally expressed genes, many of which are responsible for the onset of substantial cellular differentiation and initial organogenesis. Their expression may often, but not always, be of short duration and inde-pendent of environmental influences during this phase of embryo-genesis. It should also be realized that the same proteins may serve a different function later during postembryonic development or adult life. Hence, the cumulative expression of these genes, often encod-ing low copy number mRNAs, is depicted as a sharp narrow peak during early development in Fig. 1.2. Further differentiation leading to the expression of adult structures and functions does not often occur for a considerably longer period of time, relative to that required for early embryogenesis. This can vary from a few hours in insects to months or even years for many amphibia, reptiles and mammals. This long "pause" is characterized by considerable growth of the embryo, larva or fetus with relatively little further differentiation or activation of new genes, compared with the rapid events that are characteristic of early embryogenesis.[1,4,5,18] In free-living embryos this pause is rather abruptly terminated by the activation of genes encoding "adult" proteins and cellular components. In mammals, this transition is often gradual during gestation and continues into the perinatal period. But both cases of postembryonic development are characterized by extensive transcriptional activity, often involving the activation of genes encoding abundant mRNAs whose products have highly specialized functions and whose expression is tissue-specific. More importantly, most of these genes activated during postembryonic development continue to be expressed constitutively so that their regulation serves both to establish and maintain the adult phenotype.

Differences Between Early- and Postembryonic Developmental Processes

Although early embryogenesis and postembryonic development are both characterized by intense gene activity, there are important morphogenetic and biochemical differences between these two processes which are worth considering in the context of signaling mechanisms, some of which are listed in Table 1.2.

Most signaling molecules regulating early embryonic development are peptides which are not synthesized in specialized cells or are small molecules derived from nutrients, such as retinoic acid. Those synthesized within the early embryo are usually autocrine (acting on the same cell in which they are made) or paracrine (acting on neighboring cells) factors. The peptide growth factors (e.g., activin, bFGF and TGF-β-like) and their receptors are synthesized on maternal mRNAs stored in the egg.[12,15,21-24] The same growth factors often serve as signaling molecules later in development and in the adult organism when they are made in specialized tissues as, for example, activin and bFGF in the pituitary (see ref. 25). By far the largest group of signaling molecules synthesized upon completion of early embryogenesis are made in endocrine tissues and, hence, classified as hormones. Hormones encompass virtually every major type of large and small molecules and include such well-known

Table 1.2. Some differences between early embryogenesis and postembryonic development

Early Embryogenesis

Autonomous; not influenced by environment
Developmental signals are largely autocrine or paracrine
Signals control both determination and initial differentiation
    of undifferentiated cells
Not usually a period of sex determination
Cell death not extensive and is highly restricted spatially
Cell-cell interactions involved in the initiation of differentiation

Postembryonic Development

Strongly influenced by environmental signals
Most signals are of endocrine origin
Signals control completion of differentiation of partially differentiated cells
Onset of sexual differentiation and secondary sexual characteristics
Cell death extensive and often entails complete histolysis
    or tissue regression
Cell-cell interactions are part of the inductive response or cell death program

effectors as neuropeptides, pituitary trophic factors, prolactin, thyroid hormones, sex steroids, ecdysteroids and glucocorticoids[25,26] (see also chapter 2). An interesting feature of endocrine signals is that the hormone receptor genes are usually expressed early in development, whereas the ligands are made at later stages. In both vertebrates and invertebrates, the central nervous system plays an important role in intiating a cascade of endocrine signals, whose function is essentially to coordinate diverse functions in the developing organism in response to environmental cues. It can be generalized that early embryogenesis is autonomous since it proceeds without being significantly influenced by extra-embryonic factors, whereas postembryonic development is strongly influenced by environmental factors. One feature shared by both early and late developmental stages is that many signaling molecules, particularly growth factors and their receptors, are oncogene-derived products.[13,22,25] In postembryonic development, the different target cells of growth and developmental hormones are already partially differentiated in which the hormone initiates the completion of the process.[27] It is also quite common to find that different target cells respond to the same developmental hormone quite differently according to distinct cell-specific developmental programs[25,26] (see chapter 2).

Among other features listed in Table 1.2 that are different in the early and late developmental processes are programmed cell death, sexual differentiation and cell-cell interactions. Cell death during early embryogenesis is usually not extensive or spatially marked, and for which specific extracellular signals initiating the process have generally not been identified.[28] On the other hand, PCD during postembryonic development is usually extensive and spatially and temporally well-defined. Extracellular signals that initiate and maintain, or prevent, cell death during this period have been well-characterized, and often the same signal that induces PCD in one cell-type can promote growth and morphogenesis in another.[29-31] A good example of this dual function is hormonal regulation of metamorphosis, where thyroid hormones and ecdysteroids activate cell death and morphogenesis in different tissues of the vertebrates and invertebrates, respectively[27,32,33] (see chapters 5,8). During metamorphosis or postembryonic development, cell death is highly spatially restricted as in the restructuring of the central nervous system and gut,[34-36] or may involve the elimination of whole organs such as the tail and gills in amphibian metamorphosis and salivary gland in insect metamorphosis.

Differentiation of sex-determined tissues generally occurs during postembryonic development and is under strict hormonal or (indirectly) environmental control.[3,4,37,38] It is worth noting that in sexual dimorphism a major action of hormones is to induce cell death and regression in parallel with morphogenesis of accessory sexual tissues, as is known for Müllerian/Wolffian duct regression and oviduct or seminal vesicle development induced by estrogenic and androgenic hormones.[25] As regards cell-cell interactions, extracellular matrix (ECM) components such as cadherins and integrins are now known to act as developmental signals during early embryogenesis.[16,36,39] Although ECM molecules continue to play an important role later in life, especially in tissue remodelling, they do not act specifically as regulatory signaling molecules, but as downstream elements responding to other signals. In other words, in postembryonic and adult organisms, qualitative or quantitative changes associated with ECM components can equally be the consequence as the primary cause of the onset of a developmental process.

Table 1.3. *Some examples of the expression of 'adult' genes and processes in partially different tissues of vertebrates during postembryonic development*

Process	Genes and Tissues
Gene switching	α-Fetoprotein to albumin (liver) Fetal or larval to adult hemoglobin (hematopoietic tissues)
Morphogenesis	Components for formation of limbs, lungs; chondrogenesis
Neural development	Neuronal cell turnover; acquisition of new functional, sensory and behavioral characteristics
Tissue restructuring	Keratinization (epidermis); remodelling of digestive tract; extracellular matrix
Hormonogenesis	Activation of genes leading to hormone biosynthesis in endocrine tissues (pituitary, thyroid, gonads, etc.)
Cell death	Total or partial regression of tissues or organs by lytic enzymes (patterning of limbs, loss of tail, gills)
Sexual differentiation	Genes involved in sex determination and differentiation of accessory sexual tissues

Expression of 'Adult' Genes During Postembryonic Development

An exhaustive biochemical literature has accumulated over the last 60 years on the de novo formation of specific gene products or the establishment of novel processes during larval, fetal or neonatal development that characterizes the adult phenotype.[1,3-5,40] Some of these are listed in Table 1.3. It is important to realize that they do not represent an adaptation to new demands made upon the organism, but are the consequence of selective or de novo activation of genes in anticipation of new demands or change in environment, nutrients, etc. In recent years, many of the biochemical findings at the protein level have been extended to the expression of specific genes; for example, the switching from fetal to adult hemoglobin, α-fetoprotein to albumin, switching to adult-type visual pigmentation, skin keratinization in vertebrates or cuticle formation in insects, selective repopulation of neurons and formation of lung surfactants.[27,32,33,41-43]

The above examples point to the importance of the postembryonic developmental period for the establishment of the adult phenotype. Yet, compared with the impressive recent advances in our understanding of the molecular basis of early embryogenesis, our knowledge of the induction and regulation of 'adult' genes is at a relatively primitive stage. Much of the impressive progress in elucidating the mechanisms underlying early development is due to the availability of a number of different developmental model systems and the judicious exploitation of novel techniques of cell molecular biology and genetic manipulation. The combination of these techniques with classical genetics of *Drosophila* and *Caenorhabditis*, mechanical manipulation of frog and chick embryos and the production of transgenic mice has thus made a formidable contribution.[4,5] A similar array of models and techniques has not been available for investigations on postembryonic development, particularly in mammals. The intimate physical, biochemical and physiological association between the mother and developing fetus in mammals prevents a clear separation of maternal from fetal developmental signals and responses. A developmental system based on free-living embryos of invertebrates and non-mammalian vertebrates can, however, overcome this particular disadvantage. For this and other reasons, which will be described in detail later (see chapter 5), insect and amphibian metamorphosis offers a number of advantages for analyzing the role of extracellular signaling and the regulation of expression of 'adult' genes during postembryonic development.

References

1. Graham CF, Wareing PF, eds. Developmental Biology of Plants and Animals. Oxford: Blackwell, 1976.
2. Davidson EH. Gene Activity in Early Development, 3rd Ed, Orlando: Academic Press, 1986.
3. Browder LW, Erickson CA, Jefferey WR. Developmental Biology. Philadelphia: Saunders College Publishing, 1991.
4. Gilbert SF. Developmental Biology. Sunderland, Mass: Sinauer, 1994.
5. Müller W. Developmental Biology. Berlin: Springer, 1997.
6. Jefferey WR. Maternal RNA and the embryonic localization problem. In: Siddiqui MAQ, ed. Control of Embryonic Gene Expression. Boca Raton: CRC Press, 1983:73-114.
7. Pratt HPM, Bolton VN, Gudgeon KA. The legacy from the oocyte to the embryo. In: Porter R, Whelan J, eds. Molecular Biology of Egg Maturation. London: Pitman Books, Ciba Foundation Symp 98; 1983:197-227.
8. Tata JR. Coordinated assembly of the developing egg. BioEssays 1986; 4:197-201.
9. Gurdon JB. Gene Expression during Cell Differentiation. Oxford Biology Readers No 25, Oxford: Oxford University Press, 1973.

10. Bender W. Homeotic gene products as growth factors. Cell 1985; 43:559-60.
11. Krumlauf R. Hox genes in vertebrate development. Cell 1994; 78:191-202.
12. Jessell TM, Melton DA. Diffusible factors in vertebrate embryonic induction. Cell 1992; 68:257-70.
13. McMahon AP. A super family of putative developmental signaling molecules related to the proto-oncogene *wnt-1/int-1*. Adv Dev Biol 1992; 1:31-60.
14. Gehring WJ. Homeo boxes in the study of development. Science 1987; 236:1245-52.
15. Green JBA, Smith JC. Growth factors as morphogens: Do gradients and thresholds establish body plan? Trends Genet 1991; 7:245-50.
16. Gurdon JB. The generation of diversity and pattern of animal development. Cell 1992; 68:185-99.
17. St. Johnston D, Nusslein-Volhardt C. The origin of pattern and polarity in the *Drosophila* embryo. Cell 1992; 68:201-19.
18. Slack JMW. From egg to embryo. Regional specification in early development. Cambridge: Cambridge University Press, 1991.
19. Kirschner M, Newport J, Gerhart J. The timing of early developmental events in *Xenopus*. Trends Genet 1985; 1:41-47.
20. Stern CD, Ingham W, eds. Gastrulation. Development Suppl. 1992.
21. Smith JC. Mesoderm-inducing factors in early vertebrate development. EMBO J 1993; 12:4463-70.
22. Chapman K, Jackson S, Wilkinson D, Lunt GG, eds. Extracellular regulators of differentiation and development. London: Portland Press, 1996.
23. Klein PS, Melton DA. Hormonal regulation of embryogenesis: the formation of mesoderm in *Xenopus laevis*. Endocrin Rev 1994; 15:326-41.
24. Doniach T. Basic FGF as an inducer of neural pattern. Cell 1995; 83:1067-70.
25. Baulieu E-E, Kelly P, eds. Hormones. From Molecules to Disease. Paris: Hermann, 1990.
26. Tata JR. The action of growth and developmental hormones. Evolutionary aspects. In: Golderberger RF, Yamamoto KR, eds. Biological Regulation and Development. New York: Plenum Publishing, 1984; 3B:1-58.
27. Tata JR. Gene expression during postembryonic development: Metamorphosis as a model. Proc Indian Natn Sci Acad 1994a; B60:287-302.
28. Snow MHL. Cell death in embryonic development. In: Potten CS, ed. Perspective on Mammalian Cell Death. Oxford: Oxford University Press, 1987:202-28.
29. Tomei LD, Cope FO, eds. Apoptosis: The Molecular Basis of Cell Death. New York: Cold Spring Harbor Laboratory Press, 1991.

30. Dexter TM, Raff MC, Wyllie AH, eds. Death from inside out: the role of apoptosis in development, tissue homeostasis and malignancy. Phil Trans R Soc London B 1994:345.

31. Tata JR, Hormonal regulation of programmed cell death during amphibian metamorphosis. Biochem Cell Biol 1994b; 72:581-88.

32. Gilbert LI, Frieden E, eds. Metamorphosis: A Problem in Developmental Biology. New York: Plenum Press, 1981.

33. Gilbert LI, Tata JR, Atkinson BG, eds, Metamorphosis. Postembryonic Reprogramming of Gene Expression in Amphibian and Insect Cells. San Diego: Academic Press, 1996.

34. Burd GD. Development of the olfactory nerve in the clawed frog *Xenopus laevis*: II. Effects of hypothyroidism. J Comp Neurol 1992; 315:255-63.

35. Truman JW. Metamorphosis of the insect nervous system. In: Gilbert LI, Tata JR, Atkinson BG, eds. Metamorphosis. Postembryonic Reprogramming of Gene Expression in Amphibian and Insect Cells. San Diego: Academic Press. 1996:283-320.

36. Ishizuya-Oka A, Shimozawa A. Inductive action of epithelium on differentiation of intestinal connective tissue of *Xenopus laevis* tadpoles during metamorphosis in vitro. Cell Tissue Res 1994; 277: 427-36.

37. Galien L. Embryogénie experimentale. C R Acad Sci Paris 1953; 237:1565-66.

38. Witschi E. Development of Vertebrates. Philadelphia: WBSaunders, 1956.

39. Roskelley CD, Srebrow A, Bissell MJ. A hierarchy of ECM-mediated signaling regulates tissue-specific gene expression. Curr Opin Cell Biol 1995; 7:736-47.

40. Weber R. Biochemistry of amphibian metamorphosis. In: Weber R, ed. The Biochemistry of Animal Development. New York: Academic Press, 1967:227-301.

41. Nienhuis AW, Stamatoyannopoulos G. Hemoglobin switching. Cell 1978; 15:307-15.

42. Nahon J-L. The regulation of albumin and α-fetoprotein gene expression in mammals. Biochimie 1987; 69:445-59.

43. Weber R. Switching of globin genes during anuran metamorphosis. In: Gilbert LI, Tata JR, Atkinson BG, eds. Metamorphosis. Postembryonic Reprogramming of Gene Expression in Amphibian and Insect Cells. San Diego: Academic Press, 1996:567-97.

CHAPTER 2

Growth and Developmental Hormones and Their Actions

The term hormone (from the Greek meaning "I activate") was first introduced in 1902 by Bayliss and Starling[1] to describe a pancreatic secretion. Hormones in higher animals can be best defined as chemical messengers, produced and secreted by specialized endocrine cells to regulate the activities of their target cells. They serve to coordinate the function of diverse cell types of an organism directly or indirectly in response to environmental signals. The latter, which can be physical (light, temperature) or chemical (nutrients, olfactory or gustatory agents), are transmitted to the endocrine tissues via the central nervous system. There is virtually no metabolic growth or developmental process in multicellular organisms which is not regulated at some stage by hormones. Several textbooks and exhaustive reviews describe in detail their nature, biosynthesis and actions. The reader is referred to some of these.[2-8]

Major Characteristics of Hormones

Hormones can be classified in a number of ways: according to their chemical nature, the endocrine tissues in which they are synthesized; the regulation of their biosynthesis; their physiological actions and the biochemical and molecular processes they control. Since this monograph largely deals with how hormonal signalling regulates development, hormones are considered here in the context of their actions and the receptors through which the hormonal signals are transduced. Tables 2.1 and 2.2 list some well-known animal hormones according to whether they exert rapid metabolic effects independently of gene activity and those with the relatively slower growth and development actions that require some level of control of gene expression. The following major characteristics emerge from these two tables:

Hormonal Signaling and Postembryonic Development,
by Jamshed R. Tata. © 1998 Springer-Verlag and R.G. Landes Company.

Table 2.1. Some animal hormones that exert a rapid, metabolic action, initially independent of RNA or protein synthesis

Hormone	Chemical Nature	Major Physiological and Biochemical Actions
Epinephrine	Catecholamine	Cardiac function; thermogenesis; glycogen breakdown; regulation of cyclic AMP levels
Prostaglandins	Fatty acid derivatives	Inflammatory responses; uterine muscle activity; regulation of phosphodiesterase and cyclic AMP breakdown
Vasopressin	Nonapeptide	Water and ion transport; modulation of Na^+ pump
Growth hormone	Protein	Lipolysis; amino acid transport
Prolactin	Protein	Water and ion transport in fish and amphibia
Insulin	Protein	Sugar transport; carbohydrate metabolism; facilitates DNA synthesis

1) Hormones are not restricted to any particular class of chemicals. They range from amino acids, fatty acid derivatives, steroids, small peptides to large multimeric proteins.[4,7,8]

2) Most hormones elicit multiple biochemical actions in their target cells or tissues. It is not uncommon to find that the same hormone can rapidly regulate metabolic activity or transport of nutrients and ions, independently of protein or RNA synthesis, while also exerting a slower, growth and developmental action via control of gene expression. Good examples are insulin, growth hormone and the pituitary tropic hormones. One of the best examples of the diversity of actions of a single hormone is that of thyroid hormone.[9-11] As shown in Table 2.3, these range from such developmental actions as amphibian metamorphosis and mammalian fetal neural development to metabolic actions as maintenance of oxidative phosphorylation and thermogenesis in adult mammals.[10] Other good examples of the multiplicity of responses are the growth-promoting and lipolytic actions of growth hormone,[12,13] the rapid glucose transport and slow regulation of thyroglobulin biosynthesis in thyroid cells by thyrotropic hormone. (TSH)[14,15] and the ion transport and milk-protein gene regulatory activities of prolactin.[13,16,17]

3) There is a high degree of target-cell specificity for each hormonal response. It highlights the importance of the coupling of a common receptor to different post-receptor downstream responses

Table 2.2. Some animal hormones with relatively slow growth and developmental actions dependent on regulation of gene expression

Hormone actions	Chemical nature	Major physiological and biochemical
Ecdysone	Steroid	Invertebrate metamorphosis; gene puffing; cuticle formation; cell death in salivary gland and gut; gene transcription
Triiodothyronine	Iodoamino acid	Amphibian metamorphosis; activation of albumin and hemoglobin genes; mammalian growth, development and metabolic regulation; fetal brain development; cell death in amphibian
	metamorphosis	
Estradiol	Steroid	Growth and maturation of accessory sexual tissues and egg development; activation of egg protein genes in oviparous vertebrates; regulation of sexual dimorphism
Growth hormone	Protein	Growth and development in vertebrates; regulation of protein synthesis
Prolactin	Protein	Control of lactation and milk protein gene expression in mammals; induction of 'water drive' in terrestrial amphibia; salt metabolism in fish

which are determined at the time of differentiation of different cell-types. Perhaps the best example of the individuality of action is that of each insect and amphibian tissue responding differently to ecdysteroids and thyroid hormone during metamorphosis[18-20] (see chapter 5).

4) Metabolic activity, growth and development of a given tissue are often regulated simultaneously by the coordinated action of multiple hormones. This complexity could either be due to different cell types within a tissue responding individually to different hormones or reflect the enhancement by one hormone of the sensitivity of a given tissue to other hormones. Well-known examples of both types of such multi-hormonal cooperation are:

1) The coordinated regulation of maturation of mammary gland and lactation by prolactin, estrogen, progesterone, glucocorticoid hormone and insulin.[2,13,21,22]

2) The cooperative regulation of expression of the $\alpha_2\mu$-globulin gene in mammalian liver by androgen, thyroid hormone and growth hormone.[23,24]

3) The potentiation by progesterone and glucocorticoid hormone of the actions of estrogen on growth of mammalian uterus and egg protein gene expression in oviparous vertebrates, or by glucocorticoids of multiple actions of thyroid hormone in amphibian metamorphosis.[6,8,25]

An important question arises as to whether the multiple regulatory signals act through a unique or different pathways. It will be considered in detail for the multihormonal control of mammary gland development and lactation.

Evolution of Hormones and Their Actions

It is useful to consider some aspects of the evolution of hormones and their responses in order to evaluate the modality of hormonal regulation of growth and developmental processes. Hormones have been highly conserved through evolution. Most of the endocrine glands shown in Fig. 2.1 produce the same or similar hormones with identical functions in a wide range of vertebrates. Small molecular hormones such as thyroid hormone, glucocorticoids, sex steroids and catecholamines are the same in fish and man, as in other species (see refs. 2, 3, 26). There are some structural variations in different species for peptide and protein hormones produced in the neuroendocrine system, pancreas and digestive tract but these too have by and large retained their major functions across different species. In some cases there is retention of cross-species responsiveness to protein hormones, but the physiological action may be different, as, for example, human prolactin is equally effective in man, fish and amphibia but with different end-effects (see Tables 2.1 and 2.2).

Hormones are primitive substances on an evolutionary timescale. A commonly held view is that most hormones have arisen as by-products or waste-products of primitive metabolic processes.[26,27] This is particularly true of peptide hormones derived by partial digestion of larger proteins in the gut and tissues with specialized digestive functions such as the salivary gland and pancreas.[28] Substances which function as plant hormones (auxins), e.g., indole acetic acid, are closely related to neurotransmitters, e.g., serotonin, in higher animals. Thyroid hormones and derivatives have been detected in

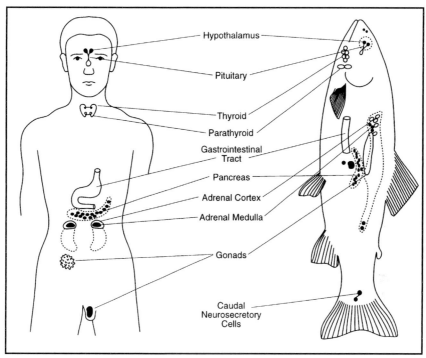

Fig. 2.1. A schematic representation of some endocrine organs in two vertebrates, spanning a long evolutionary period, which produce similar hormones, but whose functions may vary according to the species. Adapted from: Gorbman A, Bern HA. A Textbook of Comparative Endocrinology. New York: John Wiley, 1962.

primitive organisms such as algae and sponges, while the steroid hormone estradiol is present in a large amount in some primitive ferns.[26,27,29] Interestingly, the fungal steroids antheridiol and oogoniol, which regulate sexual reproduction in fungi like *Achlya*, are structurally very similar to the vertebrate male and female sex steroid hormones, testosterone and estradiol[30] (see Fig. 2.2). The catecholamines epinephrine (adrenaline) and norepinephrine are likely to be by-products of melanin or pigment formation in primitive organisms. What is important here is to realize that these substances do not always exert a hormone-like activity in lower organisms, so that they have acquired the function of chemical messengers in higher animals later in evolution. Thus the question of evolution of hormone receptors has acquired considerable importance.

Whereas the small molecular non-protein hormones mentioned above have not undergone chemical changes, there is considerable evidence that peptide hormones have undergone molecular

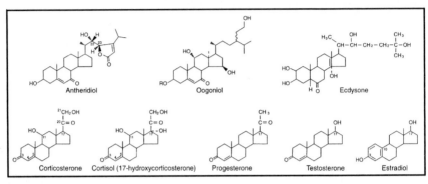

Fig. 2.2. Some steroid hormones that regulate highly specific functions in both primitive and higher organisms. These steroids include such hormonal signals as the fungal gametogenic steroids antheridiol and oogoniol, the invertebrate metamorphosis hormone ecdysone, and the vertebrate sex hormones testosterone, estradiol and progesterone. Each of these steroid hormones acts through distinct receptors encoded by different genes (see chapter 3).

evolution. Until recently, peptide hormones were thought not to occur in plants. However, in 1996 Schell's group described the structure and function of a peptide plant growth regulator.[31] Among the well-known peptide and protein hormones found in the animal kingdom are such larger protein hormones as insulin, growth hormone, prolactin, the pituitary glycoprotein hormones ACTH, TSH, LH and FSH, and the smaller (<10 amino acids) peptide hormone families of oxytocin and vasopressin.[8,27,28,32-34] There can be significant species variation in the composition and amino acid sequence of a given protein hormone resulting in a species-specificity of its responses. With the advent of cloning and sequencing of DNA, it became possible to establish with greater accuracy the evolutionary relationships among members of protein hormone families. Fig. 2.3 illustrates this feature for the multigene family comprising human growth hormone, prolactin and placental lactogen.[34] According to this scheme, chromosomal segregation of human prolactin and growth hormone occurred as far back as 4×10^8 years ago, while growth hormone and placental lactogen underwent an intrachromosomal recombination only about 10^7 years ago. An interesting finding emerging from cDNA sequencing is the demonstration that gonadotropin releasing hormone (GnRH) and prolactin inhibiting factor (PIF), synthesized in the hypothalamus, are both encoded by the same gene.[35] This finding explains why the enhanced synthesis of gonadotropin is so precisely coordinated with inhibition of that of pro-

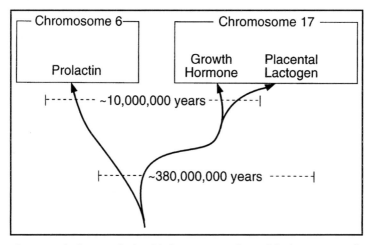

Fig. 2.3. Evolutionary relationship between members of the human growth hormone/prolactin/placental lactogen multigene family and their chromosomal segregation. Based on Cooke NE, Coit D, Shine J et al. Human prolactin. cDNA structural analysis and evolutionary comparisons. J Biol Chem 1981; 256:4007-16.

lactin in the pituitary, an essential requirement for the strict timing of changes in ovarian function during oogenesis and reproductive activity.

As regards evolution of hormonal responses, one has to consider the important characteristic of hormone action, namely the diversity of responses to a given hormone in different species or in different tissues of the same organism. Both features emerge from Table 2.3 depicting the multiplicity of actions of thyroid hormone.[9-11] Thus, in amphibia thyroid hormone regulates metamorphosis while in mammals it is important for neural maturation during fetal life and the regulation of basal metabolic rate through mitochondrial respiration. During metamorphosis each tissue of the same amphibian species responds in a dramatically different way to thyroid hormone.[19,20,36,37] The hormone induces in the tadpole morphogenesis of limb buds, gene switching in liver, remodelling of the CNS and extensive or total cell death in the tail and intestine. The same is true for insect metamorphosis where ecdysteroids control morphogenesis, gene reprogramming and cell death in different larval and pupal tissues[19,20] (see chapter 5). Other steroid hormones in vertebrates also exhibit similar diversity of actions, e.g., regulation by estrogen of egg protein genes in the liver and oviduct of oviparous vertebrates and development of accessory sexual tissues (uterus, mammary

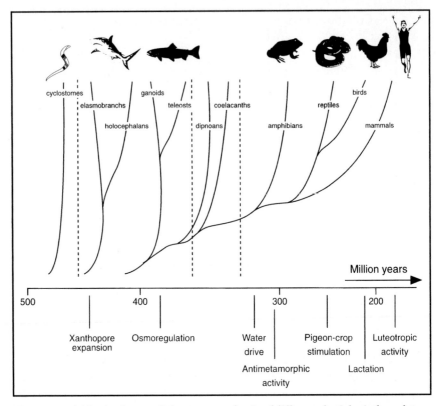

Fig. 2.4. Acquisition by the protein hormone prolactin of different physiological regulatory functions in different species acquired during evolution. At the bottom of the figure is the evolutionary time-scale, below which are given some of the important regulatory functions of prolactin. Adapted from Nicoll CS. Physiological actions of prolactin. In: Knobil E, Sawyer WH, eds, Handbook of Physiology (Section 7, Vol 4, part 2) Washington, American Physiological Society. 1974:253-92.

gland, oviduct) in mammals.[8,25,38] An appropriate example of evolutionary diversity of responses to a protein hormone is that of prolactin. As illustrated in Fig. 2.4, the major functions of this hormone in mammals are the regulation of lactation and luteotropic activity. Prolactin also regulates major functions in non-mammalian species. It stimulates crop-sac development in birds, induces "water drive" in terrestrial urodeles and regulates salt adaptation and melanogenesis in fish. Prolactins from lower vertebrates generally do not exert their function in higher forms, whereas mammalian prolactin is quite effective in fish and amphibia.[17,36,37,39] Since the amino acid sequence of prolactins is known to vary in different species, this latter fact

Table 2.3. Multiplicity of physiological and biochemical actions of thyroid hormone

Growth and Developmental actions	Metabolic actions
Rate of postnatal growth of many mammalian and avian tissues	Regulation of basal metabolic rate in homeotherms
Functional and biochemical maturation of fetal brain and bone	Movement of water and Na^+ ions across cell membranes
Morphogenesis, gene switching and cell death in amphibian larval metamorphosis	Calcium and phosphorous metabolism
Control of molting in birds	Regulation of metabolism of cholesterol and other lipids
Regulation of synthesis of mitochondrial respiratory enzymes and membranes	Nitrogen (urea, creatine) metabolism
	Control of oxidative phosphorylation and energy metabolism

reflects a concomittant evolutionary variation in the structure of protein hormone receptors as well.

Environmental Cues and Endocrine Cascades

As already mentioned, one of the most important characteristics of hormones is that they are chemical intermediaries in the transfer of information from the environment to the organism and from one group of cells to another within the individual organism. Such information transfer in higher organisms occurs as a cascade of positive and negative regulation of endocrine organs. Its overall purpose is to coordinate multiple activities of different target cells in order to facilitate the organism to adapt to environmental demands on metabolic and developmental functions.[40,41]

Fig. 2.5 presents one such example of endocrine cascades during growth and reproductive development in birds in response to various external stimuli. Most of the external signals are physical in nature, i.e., photoperiodicity, temperature, etc., which are converted into a series of nervous impulses and the release of neurotransmitters which then impinge on the neurosecretory cells of the hypothalamus. The major function of hypothalamic neurosecretory cells is to produce releasing hormones or factors such as TSH releasing hormone, corticotropin releasing factor, LH/FSH releasing hormone, prolactin inhibitory factor (TRH, CRF, LHRH, PIF, respectively) which

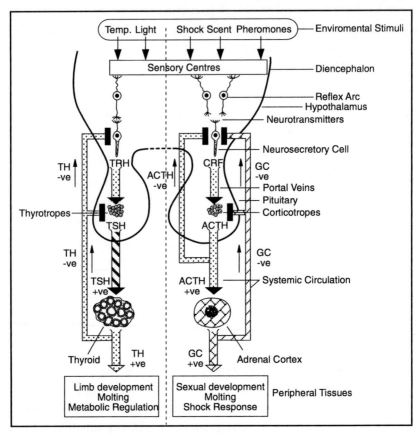

Fig. 2.5. Two examples of the conversion in the CNS of information received as physical stimuli to chemical signals giving rise to a series of endocrine cascades leading to important physiological regulatory functions. The upward and downward arrows in the cascade indicate positive (+ve) or negative (-ve) feedback. Abbreviations: TH, thyroid hormone; TSH, thyroid stimulating hormone; TRH, TSH releasing hormone; ACTH, adrenocorticotropic hormone; CRF, corticotropic releasing factor; GC, glucocorticoid hormone. TRH and CRF are produced by the hypothalamus and act on the pituitary to produce TSH and ACTH which act on the thyroid and adrenal cortex, respectively.

are transported via the portal vein to the pituitary. Each of these releasing hormones or factors, which are small peptides, acts on distinct groups of anterior pituitary cells (e.g., thyrotropes, corticotropes, gonadotropes, lactotropes, etc.) to regulate the production of corresponding tropic hormones (TSH, ACTH, LH/FSH, prolactin). These in turn act on different endocrine glands (thyroid, gonads, adrenals, etc.) to regulate the production of hormones which are released into the general circulation to reach their ultimate individual target cells.[7,15] A similar endocrine cascade has been characterized in all vertebrates and an analogous neuroendocrine network operates in invertebrates, particularly well-known for metamorphosis (see refs. 8, 19, 20). It will also be seen from Fig. 2.5 that the overall homeostasis of response is characterized by a series of negative feedback loops between the hormones produced at each step in the cascade. Although the negative feedback loop is a classical feature of endocrinology, it is worth noting that some developmental systems depend on positive feedback loops (see chapter 5).

Another example of a well-studied endocrine cascade is that operating for regulating reproductive function in oviparous animals. Hormonal control of oogenesis and egg maturation in fish, amphibia, reptiles and birds is also initiated by environmental factors such as photoperiodicity, temperature, smell, nutrients, ionic balance, etc. As illustrated in Fig. 2.6, these stimuli are converted into neurohormonal signals in the brain which, in turn, activate other endocrine organs through a cascade-type pathway. Ultimately, gonadal hormones are the key regulators of reproductive function. In oviparous female vertebrates, the major role of ovarian hormones (estrogen and progesterone) is to coordinate a number of diverse physiological and biochemical processes in different tissues which are responsible for egg maturation.[25,38,42,43] As shown in Fig. 2.6, there are three major extra-oocyte sites of synthesis of products essential for the maturing oocyte: the liver (or fat body in invertebrates), the oviduct and follicle (or nurse) cells. In vertebrates, the activity of follicle cells is under the control of pituitary gonadotropins, LH and FSH. Mammalian follicle cells are known to provide the developing egg with zona pellucida glycoproteins which play an important role in fertilization.[44] Whether non-mammalian follicle cells also make similar egg components is not known, but in all vertebrates these ovarian cells synthesize and secrete estrogen and progesterone into the bloodstream. These two hormones regulate the production of several important egg components, many essential for the early development of free-living embryos of oviparous animals in the liver and oviduct. The former is responsible for the synthesis of protein, carbohydrates and lipids that make up the yolk. The activation of genes encoding the precursor to egg yolk proteins, vitellogenin by estrogen will be discussed in detail later (see chapter 4). Estrogen also regulates the production of such key egg white components as ovalbumin, conalbumin, avidin in the avian oviduct and jelly coat proteins (e.g., FOSP-1)

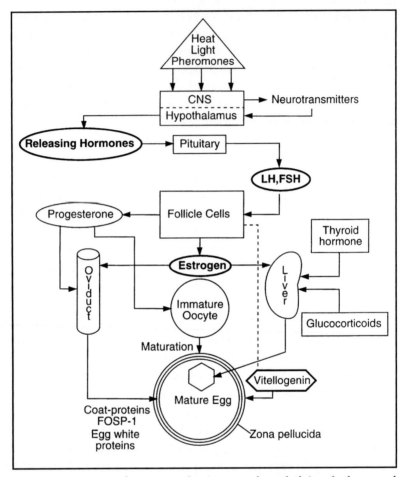

Fig. 2.6. Environmental cues in endocrine cascade underlying the hormonal regulation of egg development in oviparous vertebrates. There are three final hormonal targets: the developing oocyte, oviduct and liver, each responding differently to estrogen, which is the principal hormonal signal, with progesterone, thyroid hormone and glucocorticoids playing a facilitative role. Abbreviations: LH, FSH, gonadotropins; FOSP-1, frog oviduct-specific protein-1. See for further details: Tata JR, Lerivray H, Marsh J, Martin SC. Hormonal and developmental regulation of *Xenopus* egg protein genes. In: Roy AK, Clark JH, eds. Gene Regulation by Steroid Hormones IV, 1989:163-80.

in amphibia. Progesterone plays a secondary role while thyroid hormone and glucocorticoids are known to modulate the action of estrogen in the oviduct.[25,38,42] Since the production of gonadotropins is ultimately regulated by external stimuli via neural signals and hypothalamic releasing hormones, the overall endocrine integration serves to determine the timing and rate of egg maturation as a function of environmental stimuli.[40,41]

There are also several other examples of the interaction between the environment and endocrine system, whereby a cascade of sequential hormonal signals regulate metabolic activity, growth, development and reproduction (see ref. 41). Each hormone serves to coordinate the activities of several independent responses of individual tissues so that overall effect of the integration of hormonal signals, via environmental cues, is to ensure the timing and rate of progression of important developmental and physiological functions.

Multihormonal Control of Developmental Processes

As postembryonic developmental systems became increasingly complex, their regulation came under the control of multiple hormones. The dual complexity of a developmental process on the one hand, and that of multiple hormonal inputs on the other, is perhaps best exemplified by the regulation of maturation of the mammary gland and lactation following pregnancy.

Since the first experimental demonstration by Folley (see ref. 45 for a historical account) that lactation was dependent on ovarian hormones, numerous studies in experimental animals and humans from the laboratories of Topper, Nandi, Rosen, Houdebine and Groner have defined the regulatory interplay among a large number of hormones.[2,21,46-49] Fig. 2.7 summarizes the participation of some of the principal hormones ensuring the normal development of the mammary gland prior to and during pregnancy, as well as for initiation and maintenance of lactation following parturition. It is important to realize that each hormone acts in a sequential manner, that a given hormone may control different processes at different developmental and functional stages, and that often the optimum activity is achieved by a cooperative or synergistic interaction between two or more hormones.

As with most complex vertebrate postembryonic developmental processes, the hypothalamo-pituitary axis initiates the endocrine cascade leading to lactation (see Figs. 2.5, 2.6). This initiation is also represented at the top of Fig. 2.7. Gonadotropin releasing hormone (GnRH) and prolactin inhibiting factor (PIF) secreted by the hypothalamus play a most important role at the top of the regulatory cascade by acting on the pituitary. Increased GnRH secretion activates the formation and release of gonadotropin (LH, FSH), while the reduction in secretion of negatively acting PIF enhances the production of prolactin (PRL) by the pituitary. PRL is crucial for the

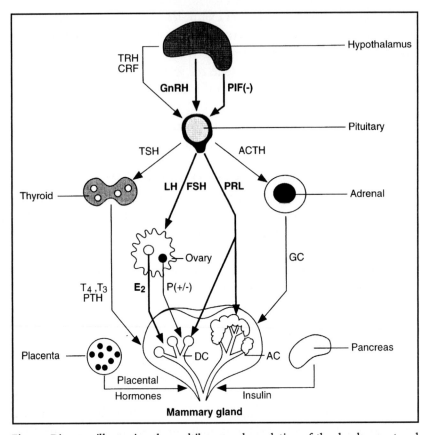

Fig. 2.7. Diagram illustrating the multihormonal regulation of the development and lactational function of the mammary gland. The hormones that play a major role are in bold, while the others serve a facilitative or supportive function. All hormones have a positive or stimulatory action, except prolactin inhibitory factor (PIF) which suppresses prolactin (PRL) synthesis and secretion by the pituitary. Progesterone (P) can either exert a stimulatory or inhibitory effect depending on the celltype and developmental stage. Estrogen (E$_2$) and PRL are the principal hormones which first control major steps in development and later the production and secretion of milk protein, carbohydrate and fat by alveolar (AC) and ductal (DC) cells of the mammary gland. Glucocorticoid (GC), insulin, thyroid hormones (T$_4$, T$_3$), parathyroid hormone (PTH) and placental hormones all play a secondary, but important, role at different developmental stages in the functioning mammary gland. Other abbreviations as in Fig. 2.5.

initiation and maintenance of lactation, whereas estrogen (E$_2$) produced by the ovary in response to elevated gonadotropin levels is vital for the development of alveolar and ductal tissues at the onset and during pregnancy. Progesterone (P) acts antagonistically with respect to E$_2$ at certain stages by inhibiting or diminishing lactation. Glucocorticoid hormone (GC), insulin, thyroid (T$_4$, T$_3$), parathyroid

(PTH) and placental hormones (especially placental lactogen) all exert important facilitative or supportive actions through different pathways for the functional maturation of the mammary gland and maintenance of lactation. For example, at the ultrastructural level, insulin promotes the proliferation of alveolar cells and hydrocortisone enhances the formation of the secretory apparatus of these cells, in advance of the stimulatory action of prolactin on the synthesis of milk proteins.[47]

The advent of gene cloning and recent immunochemical technologies have offered suitable gene and protein markers, thus making it easier to dissect the role of each of the members of the multihormonal complex. Particularly important in this context has been the cloning of genes encoding the milk proteins caseins and lactalbumin. Rosen's group have clarified some of the important mechanisms by which prolactin exerts transcriptional and post-transcriptional control over casein gene expression by studying it in mammary gland organ culture.[49,50] While the addiction of PRL to these cultures rapidly enhances the rate of transcription of casein genes, the long-term build-up of the mRNA is largely due to a selective multihormonal stabilization of casein transcripts. This dual transcriptional and post-transcriptional potentiation of specifically induced mRNA is a common feature of the hormonal induction of specialized gene products, as for example, the induction by estrogen of egg white proteins and vitellogenin in the oviduct and livers, respectively, of oviparous vertebrates[25,43,51] (see chapter 4). In the case of complex multihormonal regulation of lactogenesis, it is not enough to enhance the accumulation of milk protein mRNAs, but that a number of other processes necessary to sustain lactation and dependent on the control by other hormones (Fig. 2.7) have to be coordinated temporally and spatially. It is this integrative endocrine control that is the key to normal development and function of specialized tissues such as the mammary gland. Hormonal integration, in turn, first involves a specific interaction between each individual hormone and its receptor and secondly a cross-regulation of receptors by the different hormones that constitute a multihormonally regulated developmental and physiological process. Both these aspects of hormone action are dealt with in chapters 3 and 7.

References

1. Bayliss WM, Starling EH. The mechanism of pancreatic secretion. J Physiol 1902; 28:325-53.
2. Gorbman A, Bern HA. A Textbook of Comparative Endocrinology. New York: John Wiley, 1962.
3. Barrington EJW. An Introduction to General and Comparative Endocrinology. Oxford: Clarendon Press, 1963.
4. Turner CD, Bagnara J. General Endocrinology. Philadelphia: Saunders, 1971.
5. Gaillard PJ, Boer HH, eds. Comparative Endocrinology. Amsterdam: Elsevier/North Holland, 1978.
6. Tata JR. The action of growth and developmental hormones. Biol Rev 1980; 55:285-319.
7. Norman AW, Litwack G. Hormones. Orlando: Academic Press, 1987.
8. Baulieu E-E, Kelly PA, eds. Hormones. From Molecules to Disease. Paris: Hermann, 1990.
9. Pitt-Rivers R, Tata JR. The Thyroid Hormones. London: Pergamon Press, 1959.
10. Tata JR. Growth and developmental actions of thyroid hormones at the cellular level. In: Handbook of Physiology, Section 7, Endocrinology 3. Washington DC: American Physiological Society, 1974:469-78.
11. De Groot LJ, Larsen PR, Hennemann G. The Thyroid and its Diseases. New York: Churchill Livingstone, 1996.
12. Scharf J-H, ed. Das Somatotrope Hormon. Deutsche Akademie der Naturforscher Leopoldina, Halle/Saale, 1974.
13. Kelly PA. Growth hormone and prolactin. In: Baulieu EE, Kelly PA, eds. Hormones from Molecules to Disease. Paris: Hermann, 1990:191-217.
14. Dumont JE. The action of thyrotropin on thyroid metabolism. Vitamins and Hormones 1971; 29:287-412.
15. Labrie F. Glycoprotein hormones: Gonadotropins and thyrotropin. In: Baulieu EE, Kelly PA, eds. Hormones from Molecules to Disease. Paris: Hermann, 1990:257-74.
16. Nicoll CS, Bern HA. On the action of prolactin among the vertebrates: is there a common denominator? In: Wolstenholme GEW, Knight J, eds. Lactogenic Hormones. London: Churchill Livingstone, 1972:299-317.
17. Nicoll CS. Physiological actions of prolactin. In: Knobil E, Sawyer WH, eds. Handbook of Physiology. Washington: American Physiological Society, 1974; 7(4)2:253-92.
18. Weber R. Biochemistry of amphibian metamorphosis. In: Weber R. The Biochemistry of Animal Development. New York: Academic Press, 1967:227-301.
19. Gilbert LI, Frieden E, eds. Metamorphosis: A Problem in Developmental Biology. New York: Plenum Press, 1981.

20. Gilbert LI, Tata JR, Atkinson BG, eds. Metamorphosis. Postembryonic Reprogramming of Gene Expression in Amphibian and Insect Cells. San Diego: Academic Press, 1996.

21. Topper YJ. Multiple hormone interactions in the development of mammary gland in vitro. Recent Progr Horm Res 1970; 26:287-308.

22. Matusik RJ, Rosen JM. Prolactin induction of casein mRNA in organ culture. A model system for studying peptide hormone regulation of gene expression. J Biol Chem 1978; 253:2343-47.

23. Kurtz DT, Feigelson P. Multihormonal control of the messenger RNA for the hepatic protein α_{2p}-globulin. In: Litwack G, ed. Biochemical Actions of Hormones. New York: Academic Press, 1978; 5:433-55.

24. Roy AK, Sarkar FH, Murty CVR et al. Intra- and intercellular aspects of the hormonal regulation of α_{2p}-globulin gene expression. In: Roy AK, Clark JH, eds. Gene Regulation by Steroid Hormones III. New York: Springer, 1987.

25. Tata JR, Smith DF. Vitellogenesis: A versatile model for hormonal regulation of gene expression. Recent Prog Horm Res 1979; 35:47-95.

26. Barrington EJW. Hormones and Evolution. London: English Universities Press, 1964.

27. Tata JR. Evolution of hormones and their actions. Chemica Scripta 1986; 26B:179-90.

28. Mutt V. Questions answered and raised by work on the chemistry of gastrointestinal and cerebrogastrointestinal hormonal polypeptides. Chemica Scripta 1986; 26B:191-207.

29. Barrington EJW. Chemical communication. Proc R Soc London Ser B 1977; 199:361-75.

30. Timberlake WE, Orr WC. Steriod hormone regulation of sexual reproduction in Achlya. In: Goldberger RF, Yamamoto KR, eds. Biological Regulation and Development. New York: Plenum Press, 1984; 3B:255-83.

31. van de Sande K, Pawlowski K, Czaja I et al. Modification of phytohormone response by a peptide encoded by *ENOD40* of legumes and a nonlegume. Science 1996; 273:370-3.

32. Acher R. Molecular evolution of biologically active polypeptides. Proc R Soc London Ser B 1980; 210:21-43.

33. Wallis M. The molecular evolution of pituitary hormones. Biol Rev 1975; 50:35-98.

34. Cooke NE, Coit D, Shine J et al. Human prolactin. cDNA structural analysis and evolutionary comparisons. J Biol Chem 1981; 256:4007-16.

35. Nikolics K, Mason AJ, Szonyi E et al. A prolactin-inhibiting factor within the precursor for human gonadotropin-releasing hormone. Nature 1985; 316:511-17.

36. Tata JR. Gene expression during metamorphosis: An ideal model for postembryonic development. BioEssays 1993; 15:239-48.

37. Tata, JR. Hormonal interplay and thyroid hormone receptor expression during amphibian metamorphosis. In: Gilbert LI, Tata JR, Atkinson BG, eds. Metamorphosis. Postembryonic Reprogramming

of Gene Expression in Amphibian and Insect Cells. San Diego: Academic Press, 1996:465-503.

38. O'Malley BW, Tsai M-J, Tsai SY et al. Regulation of gene expression in chick oviduct. Cold Spring Harbor Symp Quant Biol 1978; 42:605-15.

39. Denver RJ. Neuroendocrine control of amphibian metamorphosis. In: Gilbert LI, Tata JR, Atkinson BG, eds. Metamorphosis. Postembryonic Reprogramming of Gene Expression in Amphibian and Insect Cells. San Diego: Academic Press, 1996:433-64.

40. Benson GK, Phillips JG. Hormones and the environment. Cambridge: Cambridge University Press, 1970.

41. Follett BK, Ishii S, Chandola A. The endocrine system and the environment. Japan Sci Soc Press, Berlin: Springer Verlag, 1985.

42. Tata JR, Lerivray H, Marsh J, Martin SC. Hormonal and developmental regulation of *Xenopus* egg protein genes. In: Roy AK, Clark JH, eds. Gene Regulation by Steroid Hormones IV, 1989:163-80.

43. Wahli W, Dawid IB, Ryffel GU et al. Vitellogenesis and the vitellogenin gene family. Science 1981; 212:298-304.

44. Wassarman PM. Profile of a mammalian sperm receptor. Development 1990; 108:1-17.

45. Folley SJ. The Physiology and Biochemistry of Lactation. Edinburgh: Oliver and Boyd, 1956.

46. Nandi S. Endocrine control of mammary gland development and function in the C3H/HeCrgl mouse. J Natl Cancer Inst 1958; 21:1039-63.

47. Mills ES, Topper YJ. Some ultrastructural effects of insulin, hydrocortisone, and prolactin on mammary gland explants. J Cell Biol 1970; 44:310-28.

48. Talamantes F Jr. Comparative study of the occurrence of placental prolactin among mammals. Gen Comp Endocrinol 1975; 27:115-21.

49. Rosen JM, Rodgets JR, Couch CH et al. Multihormonal regulation of milk protein gene expression. Annals New York Acad Sci 1986; 478:63-76.

50. Guyette WA, Matusik RJ, Rosen JM. Prolactin-mediated transcriptional and post-transcriptional control of casein gene expression. Cell 1979; 17:1013-23.

51. Shapiro DJ, Barton MC, McKearin DM et al. Estrogen regulation of gene transcription and mRNA stability. Recent Progr Horm Res 1989; 45:29-59.

CHAPTER 3

Hormone Receptors

The most important element in the immediate chain of events that initiate the action of a given hormone is the receptor. It is also true for neurotransmitters, drugs and virtually every biologically active substance. The basic principle of specific receptors to mediate the action of an active agent (ligand) arose from pharmacological studies in the 1940s and has since become the central dogma for explaining major phenomena of physiology, endocrinology and neurobiology. Not surprisingly, an enormous literature has accumulated over the last three decades and the reader is referred to several detailed books and reviews on hormone receptors (see refs. 1-6).

Three important technical developments have largely contributed to the dramatic advances made in recent years in our understanding of the structure, function and cell biology of receptors for hormones and other extracellular signaling molecules: 1) The availability in the early 1960s of radioactive hormones of very high specific activity enabled investigators to distinguish between receptor-ligand interactions of high affinity from the non-specific binding to various cellular components; 2) The advent of gene cloning and sequencing techniques, along with high resolution X-ray crystallography, gave an insight into the protein structure of the receptor and allowed one to discern the interrelationships among different classes of receptors; and 3) The advances in the study of oncogenes and their role in cellular signaling mechanisms enhanced our understanding of the function of hormone receptors and their classification. These developments have led to the concept of two major classes of hormone receptors based on their subcellular location and, to a lesser extent, on the nature of their ligand: 1) cell surface membrane receptors; and 2) nuclear receptors. Table 3.1 lists some membrane and nuclear receptors for hormonal and non-hormonal extracellular signals.

Hormonal Signaling and Postembryonic Development,
by Jamshed R. Tata. © 1998 Springer-Verlag and R.G. Landes Company.

Membrane Receptors

It is logical to expect that an important site for the sensing of extracellular signals and their transduction to intracellular regulatory machinery should reside in the target cell membrane. Consequently, considerable information had accumulated by the 1970s, based on radioligand and immunocytochemical analyses, on the localization and characterization of receptors in the cell membrane of protein and small water-soluble hormones, such as insulin, growth hormone and catecholamines.[1-5] However, it is the molecular cloning and sequencing of receptors for growth factors, such as epidermal growth factor (EGF) and platelet-derived growth factor (PDGF), that led the way to the breakthrough in our current understanding of the structure and function of membrane receptors for hormones.[7-14] The first studies soon revealed three important facets of membrane receptors, namely that they have similar modular structures, that most of them are oncogene products related to v-*erb*B (and later to other oncogenes), and that they share common features of signal transduction linked to intracellular signaling mechanisms.

Fig. 3.1 depicts a simplified consensus structure for hormone receptors in cell membranes. The three receptors illustrated exhibit both similarities and differences. Three major domains common to all membrane receptors are an extracellular ligand-binding domain, a transmembrane region and a cytoplasmic domain responsible for effecting the cellular action of the hormone. Both EGFR and IR have two extracellular subunits, but only in IR are they SH-linked. The oncogene products v-*erb*B and v-*ros* are related to EGFR and IR, while v-*mpl* is to PRLR. The extracellular and cytoplasmic domains of all receptors are linked by a transmembrane domain which can traverse the membrane with only one span, as for receptors for growth hormone and PRL[12] or multiple spans IR, EGFR and most receptors for growth factors.[7,9-11] The β-subunits of IR are linked to the α-subunits just above the transmembrane domains. The cytoplasmic domain comprises the regulatory region, which in all receptors and related oncogene products, is characterized by protein phosphorylation sites, the most important of which is a tyrosine site. In response to the hormonal signal, the cytoplasmic kinase itself may undergo autophosphorylation or catalyze the phosphorylation of a separate substrate. Tyrosine kinases play a crucial, but not exclusive, role in the signal transduction pathway leading to hormonal action.[15]

It is now over 40 years since Sutherland described how two hormones, glucagon and epinephrine, much before they were shown to

Table 3.1 *Some cell membrane and nuclear receptors for hormonal and non-hormonal extracellular signals*

| Ligand | Membrane Receptors | | Ligand | Nuclear Receptors | |
	Nature	Hormonal or non-hormonal		Nature	Hormonal or non-hormonal
Insulin	Protein	Hormone	Glucocorticoid	Steroid	Hormone
Growth hormone	Protein	Hormone	Estradiol	Steroid	Hormone
Prolactin	Protein	Hormone	Progesterone	Steroid	Hormone
TRH, CRF	Small peptides	Hormone	Testosterone	Steroid	Hormone
TSH, ACTH	Proteins	Hormone	3,3',5-Triiodothyronine	Iodo-amino acid	Hormone
Epinephrine	Catecholamine	Neurotransmitter	All-trans retinoic acid	Retinoid	Vitamin derivative
Acetylcholine	Methylamine	Neurotransmitter	9-cis retinoic acid	Retinoid	Vitamin derivative
Glutamate	Amino acid	Neurotransmitter	Vitamin D_3	Steroid	Vitamin derivative
Endorphin	Protein	Opiate/Neurotransmitter			
Epidermal Growth Factor	Protein	Growth Factor			

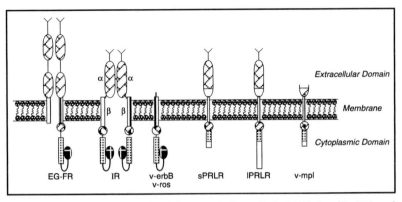

Fig. 3.1. Schematic representation of receptors for EGF (EGF-R), insulin (IR) and the short and long forms for PRL (sPRLR, lPRLR), and viral oncogene products related to the receptors (v-*erb*B, v-*ros* and v-mpl). Y, Hormone-binding region; ○, extracellular domain; ■, kinase activity; J, hinge region; ●, regulatory region containing phosphorylation site. See text for detailed explanation.

act through membrane receptors, brought about their physiological action by regulating the intracellular concentration of the second messenger, cyclic AMP (see ref. 16). The importance of this regulatory pathway became clear when adenyl cyclase was found to be physically closely associated in the cell membrane with the receptor of the hormone that regulates its activity.[1,3] In the last 20 years several key second messenger systems which are important for hormones acting via cell membrane, such as inositol trisphosphate (IP$_3$), Ca^{2+}, G protein and nitric oxide (NO) have now been described.[17-23] Fig. 3.2 illustrates typical associations between the second messengers and membrane receptors for three hormones: epinephrine, insulin and prolactin.

A major functional feature of the membrane receptors shown in Fig. 3.2 is the rapidity of response of the second messenger systems upon activation of the receptor by its ligand. Since the first description of epinephrine-activated cyclic AMP (cAMP) synthesis in the late 1950s (see ref. 16), the structure of the catecholamine receptors, the adenylate cyclase system and the multiple hormonal actions downstream of cAMP have been well established.[18] Epinephrine is the ligand for α- and β-adrenergic receptors, both associated with GTP-linked G proteins. The α receptor interacts with the stimulatory form of G protein, to generate inositol trisphosphate (which regulates intracellular Ca^{2+}) and diacylglycerol by the activation of membrane phospholipase C. The β-adrenergic receptor is linked to both the stimulatory and inhibitory forms of G protein which, via adenylate cyclase, regulate cAMP generation both positively and

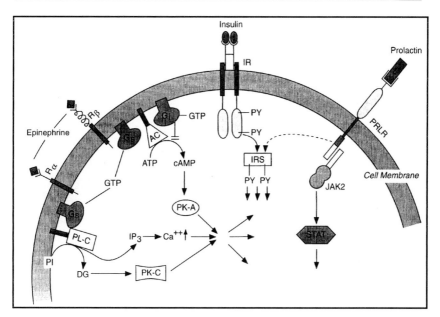

Fig. 3.2. A diagram to illustrate second messenger systems associated with membrane receptors for epinephrine, insulin and prolactin. L, ligand; Rα, Rβ, α- and β-adrenergic receptor; IR, insulin receptor; G$_s$, G$_i$, stimulatory and inhibitory G protein; PL-C phospholipase C; AC, adenylate cyclase; PI, phosphatidyl inositide; IP$_3$, inositol trisphosphate; DG, diacylglycerol; PK-A, PK-C, phosphokinases A and C; cAMP, cyclic AMP; PY, phosphotyrosine; IRS, insulin receptor substrate protein; JAK, Janus kinase; STAT, signal transduction and transcription factor. See text for explanation.

negatively. The cAMP, IP$_3$ and DG signaling systems are put to diverse uses in cellular regulation. The insulin receptor undergoes auto-tyrosine phosphorylation upon ligand activation, which in turn, regulates tyrosine phosphorylation of insulin receptor substrate protein (IRS). The consequence of the latter modification leads to several multiple downstream cytoplasmic and nuclear actions,[11] (see Fig. 3.3). Similarly, the ligand-activated prolactin receptor, which is a cytokine receptor-type, can exert multiple regulatory actions, including modulation of transcription in the cell nucleus by virtue of its modulation of the JAK/STAT intracellular signaling pathway.[12,24,25]

An example of the diversity of downstream intracellular anabolic regulatory processes triggered off by modulation of second messengers or protein phosphorylation is illustrated in Fig. 3.3 for the multiple actions of insulin and EGF. As a result of a chain of protein phosphorylation and dephosphorylation triggered by the ligand-activated receptor,[15] the EGF receptor is thought to interact with the GTP-linked *Ras* oncogene product. This in turn will activate the *Raf*-1 and IRS, a key component leading via a series of steps

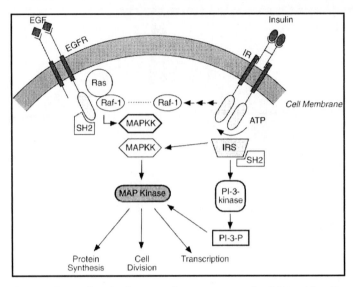

Fig. 3.3. Scheme showing how membrane receptors for EGF and insulin can exert their growth-promoting actions through the MAP kinase cascade acting at both the cytoplasmic and nuclear sites. Abbreviations: MAP, mitogen-activated protein; SH2, *src* homology domain 2. Ras and Raf-1 are oncogenic protein kinases; IRS, insulin receptor substrate protein; MAPKK, MAP kinase kinase; PI-3, kinase. Other abbreviations as in Figs. 3.1 and 3.2. See text for details.

to the activation of MAP kinase and eventually MAPK itself. The MAPK cascade is now believed to be the principal pathway for activating the protein synthetic, transcriptional and DNA synthetic machineries of all eukaryotic cells.[26] Similarly, insulin receptor will activate the MAPK cascade via *Raf*-1 stimulation of MAPK kinase. Not surprisingly, MAPK and the protein phosphorylation-dephosphorylation cascade have been found to be highly conserved through evolution.[27] However, it is important to note that the insulin signal via its receptor will be transduced via multiple intracellular pathways, depending on the cell-type, the impingement of other signals, extracellular factors, etc. Fig. 3.3 illustrates this feature for the phosphorylated IRS activating MAPK via *Raf*-1 and also via PI$_3$-kinase. IRS protein itself can be modified by tyrosine phosphorylation via the JAK-3 system regulated by other receptors, such as those of cytokines.[11,26-28] Thus it is clear that insulin (and other hormones and growth factors) can elicit its pleiotropic responses by activating its receptor tyrosine kinase to phosphorylate several intracellular proteins. Recently, phosphotyrosine residues have been shown to bind specifically to proteins that contain *Src* homology 2 (SH2) domains, so that this interaction can mediate the regulation of multiple intracellular signaling pathways.[11,28,29]

A less indirect pathway for hormonal and other signals acting through membrane receptors to regulate DNA synthesis, transcription and growth-promoting cellular functions is via the JAK/STAT and cyclic AMP-activated CREB and CREM systems. An example of hormonally regulated JAK/STAT pathway is illustrated in a highly simplified manner in Fig. 3.4 for prolactin. There is good evidence now that not only is the prolactin receptor (PRLR) structurally similar to cytokine receptors, but that major intracellular functional elements resemble those that are important for the regulation of transcription and DNA synthesis.[12,25,30] This evidence is based on the action of PRL on a classical hormonal target, the mammary gland, and a less well-known one, the T-lymphocyte. There are at least two pathways from PRLR: one following the *Raf*MEK or MAPKK-MAPK route (Fig. 3.3) and the other via *fyn*/JAK2 activating the p91 (STAT 91), which is known to be a transcription factor involved in regulation of gene expression by cytokines and interferon.[24,31,32] What is important in the context of evolution is that membrane receptors and downstream cascades leading to growth and metabolic control involves complex networks of oncogene products.[14]

Another well-documented system of regulation of gene expression by hormonal signals acting via membrane receptors is through second messengers generated by the hormone, particularly cyclic AMP.[33-35] As shown in Fig. 3.5, cAMP has been found to regulate the transcription of a number of genes through a specific DNA sequence in their promoter termed cAMP response element (CRE) which is recognized by a family of transcription factors (CREB). The ability of CREB to stimulate transcription is dependent on its phosphorylation by the protein kinase PKA, whose activity is under cAMP control. PKA and other kinases can also regulate the function of other transcription factors. Among the many genes whose promoters have CRE sequences is c-*fos*, which together with c-*jun*, is part of the DNA-binding AP-1 protein family.[34] AP-1 is now known to be induced by many growth factors, cytokines, neurotransmitters and hormones. In view of the importance of the c-*fos*/c-*jun* heterodimer in transcriptional regulation, it is significant that some of these signals act through membrane receptors that act via the JAK/STAT pathway involving serum response factor (SRF) which binds to serum response

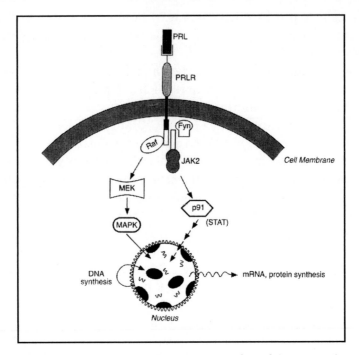

Fig. 3.4. An example of how a hormone acting through its receptor in the cell membrane can regulate nuclear functions, such as DNA synthesis and gene transcription. Prolactin-activated PRLR can either work through the Raf-MEK-MAPK pathway (Fig. 3.3) to control DNA synthesis or through the Fyn-JAK2 pathway to phosphorylate the transcription factor p91 (STAT91) and thereby control gene expression. Abbreviations as in Figs. 3.1-3.3.

element in the promoter of the c-*fos* gene. Thus two different signals acting through their membrane receptors can function cooperatively to modulate transcription. Since nuclear hormone receptors act as transcription factors, the modulation of their function by cAMP regulated protein kinases means that hormones acting through membrane receptors can modify the action of hormones acting through nuclear receptors. This relationship is often termed 'cross-talk' and will be discussed below.

Nuclear Receptors

Early Studies

The story of nuclear receptors can be said to truly begin with the synthesis of high specific activity radioactive estradiol-17β at the

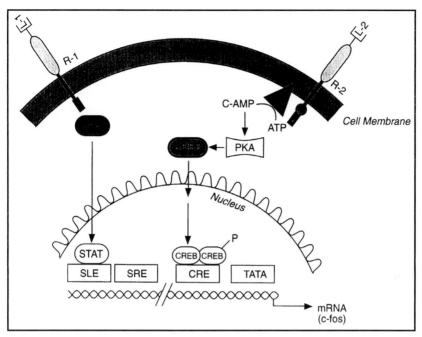

Fig. 3.5. Scheme showing how two different ligands acting through separate receptors and through the cAMP and JAK-STAT pathways can act cooperatively to control the transcription of c-*fos*, one of the important members of the c-fos/c-jun family of AP-1 transcription factors. Abbreviations: CRE, cAMP responsive element; CREB, CRE binding protein; CREB-P, phosphorylated CREB; SRE, serum response element; SRF, serum response factor. Other abbreviations as in Fig. 3.2. R-1 and R-2 are receptors for ligands L-1 and L-2.

beginning of the 1960s by Jensen and Jacobson.[36] When the ^{3}H-labeled hormone was injected into rats, it was retained for several hours in accessory female sexual tissues such as the uterus and vagina, while it was rapidly cleared from blood and non-target tissues. In later studies from Jensen's laboratory involving subcellular fractionation of the target tissues, ^{3}H-estradiol was found to be largely localized in the nuclear fraction. However, in ovariectomized animals, a substantial amount of the radioactivity was recovered in the cytosolic fraction from which many investigators attempted to purify the receptor in the earlier studies.[37-39] Upon administration of unlabeled hormone or its active analogs, the radioactive estrogen was rapidly translocated to the nuclear compartment which led Jensen to propose the classical two-step model of steroid hormone action.[40]

An important observation concerning the phenomenon of cy-toplasmic-nuclear translocation of receptor was a shift in the sedi-mentation constant of the protein-steroid complex as it moved from one cellular compartment to the other. Both the cytoplasmic and nuclear forms of the protein complexes satisfied the major require-ments of their behaving as receptors, e.g., high affinity, low capacity, ligand and tissue specificity, etc.[1,3] Similar findings were made with the other steroid hormones, namely glucocorticoid, progesterone and androgen interacting with cytoplasmic and nuclear fractions of their target tissues.[40-43] The application of techniques of autoradiography and monoclonal antibodies further confirmed the nuclear localiza-tion of steroid receptors.[44,45] Work on nuclear localization of steroid hormones received a major boost, when at about the same time, sev-eral laboratories reported that steroid and other growth promoting hormones rapidly stimulated the overall transcriptional activity of their target tissues (see chapter 4).

According to Jensen's 'two-step' model, the free steroid hormone somehow enters the cell and binds to the cytoplasmic receptor in which it would induce a conformational change detectable as a shift in its sedimentation constant. This complex is thought to enter the nucleus with a further conformational change characterized by a different response to a shift in the ionic strength of the medium. The liganded receptor would then firmly bind to the chromatin in the nucleus (see Fig. 3.6). The question of how an estrogen receptor (ER) is translocated from the cytoplasmic to the nuclear compartment was not answered until it was later established by several groups that the liganded ER formed a complex with the stress or heat-shock pro-tein hsp 90, from which it would be dissociated in the nucleus. This model explained the changes seen earlier in sedimentation constants in the two cellular compartments.[3,46,47] More recently, Milgrom's labo-ratory have obtained evidence showing that the intracellular traf-ficking of steroid hormone receptors is a two-way process, which based on receptor phosphorylation and dephosphorylation, allows a recycling of the receptor from the nucleus back to the cytoplasm.[48]

Later Studies on Structure and Function of Nuclear Receptors

As with membrane receptors, it was the cDNA cloning and se-quencing of steroid and other receptors in the mid-1980s that has contributed largely to our current knowledge of nuclear receptors. Among the first whose structures were thus established were hu-

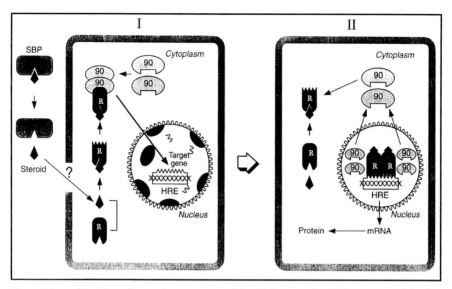

Fig. 3.6. A combination of Jensen's two-step model and chaperoning by hsp90 to explain the translocation into the nucleus of receptors of steroid hormones.
I: The extracellular steroid released from its circulating steroid-binding protein (SBP) is transported into the target cell's cytoplasm by passive diffusion or active transport. When bound to the unliganded receptor (R) it induces a conformation change allowing it to bind the heat-shock protein 90 (hsp90) dimer which acts as its chaperone. Because of the nuclear localization signal (NLS) of the receptor, the R-hsp90 complex is translocated into the nucleus. II: Once in the nucleus, the liganded receptor dissociates from hsp90 and itself dimerizes. The removal of hsp90 unmasks the DNA-binding site of the receptor, which allows the receptor to interact with the hormone responsive element (HRE) in the target gene promoter, thus activating its transcription. This mechanism is only applicable to Group I receptors, i.e., ER, PR, GR, MR, which do not heterodimerize with other nuclear receptors (see Table 3.2). The receptors of the other group reside predominantly in chromatin.

man glucocorticoid and estrogen receptors, which also revealed that they were related to the viral erythroblastosis oncogene v-*erb*A.[6,49-53] In 1986, the laboratories of Vennstrom and Evans simultaneously reported that human and chicken thyroid hormone receptors, encoded by two genes termed α and β, is the cellular homolog of this oncogene.[54,55] It was soon recognized that receptors for all steroid hormones and those of other nuclear receptors such as those for progesterone, vitamin D and retinoids, were all closely related and belong to a large supergene family commonly called the steroid/thyroid hormone/retinoic acid nuclear receptor family.[6,53,56-59] Table 3.2 lists the major members of this family. The ligands of almost all nuclear receptors are lipophilic in nature and do not offer the basis

Table 3.2. *Some members of the nuclear steroid/thyroid hormone/retinoid receptor supergene family classified according to their binding to heat-shock protein 90 and their activity as monomers, homo- or heterodimers*

Receptor	Natural ligand	Complex with hsp90	Active as	Multiple	
			Monomer or Homodimer	Heterodimer	Isoforms
Group I					
ER	Estrogens	+	+	-	-
GR	Glucocorticoids	+	+	-	-
AR	Androgens	+	+	-	-
PR	Progesterone	+	+	-	-
MR	Mineralocorticoids	+	+	-	-
Group II					
TR	Thyroid hormone	-	-	+	+
RAR	Retinoic acid	-	-	+	+
RXR	9-cis-retinoic acid	-	+	+	+
EcR	Ecdysteroids	-	-	+	+
VDR	Vitamin D_3	-	-	+	+
COUP-tf	"Orphan"	-	+	+	+
PPAR	Fatty acids	-	-	+	+

The receptors are listed as commonly used abbreviations and are identified by their natural ligands, except for COUP-tf and other "orphan" receptors for which the ligands have not been identified.

for their sub-classification. A major consequence of the molecular cloning of the nuclear steroid receptor supergene family has been the discovery of 'orphan' receptors. These are members of this gene family whose natural ligands are unknown, or were unknown when they were first characterized.[60] Most prominent among these is the retinoid X receptor (RXR), whose ligand in vertebrates was later identified as the vitamin A metabolite 9-cis-retinoic acid[61-63] and its invertebrate homolog the product of *ultraspiracle* (*usp*) gene.[64-67] Also well known orphan receptors are COUP-tf (chicken ovalbumin upstream promoter transcription factor)[60,68] and peroxisome proliferator activator receptor (PPAR).[69,70] Interestingly, one of the striking characteristics of these orphan receptors is their ability to strongly heterodimerize with other members of the nuclear receptor family.

Thanks to intensive studies over the last 15 years, much is now known about the interaction of these intracellular receptors with DNA. This interaction has led to their being divided into sub-groups, most commonly according to whether or not they interact with their target gene promoters as monomers, homodimers or heterodimers. The group comprising receptors for estrogen, progesterone, androgen and glucocorticoids interact with their response elements in the promoters as monomers or homodimers.[71,72] It is also this sub-group of steroid receptors whose translocation into the nucleus is facilitated by the hsp90 chaperones. The other, and larger, sub-group comprises receptors for retinoids, thyroid hormone, vitamin D_3 and ecdysteroids and are all known to strongly interact with their response elements as heterodimers.[58,62,73,74] Another characteristic of this sub-group is that the receptors exhibit multiple isoforms which are either products of multiple genes or generated by alternate processing of mRNA, although a novel isoform β of ER has been recently discovered.[124] In most cases the differences reside in the N-terminal part of the receptor molecule (see below). It has been suggested that this multiplicity may underlie tissue-, gene- and developmental-specificity of the multiple actions of the relevant hormone or ligand.[74-79] However, there is also evidence from gene 'knockout' (targeted disruption via homologous recombination) studies with RARs of redundancy and that more than one gene encoding the isoforms undergoing -/- mutations to obtain a strong phenotype.[80]

As the results of the early investigations on cloned nuclear receptors began to accumulate, a common modular structure soon became obvious which also established that nuclear receptors are ligand-activated transcription factors. The first receptors whose structures were thus characterized were those of estrogen and glucocorticoids.[51,81] Fig. 3.7 depicts the major structural and functional domains of several members of the steroid/thyroid hormone/retinoid receptor family. Most important are the C-terminal ligand-binding domain (LBD) (domain E) and the DNA-binding domain (DBD). Not surprisingly, the LBD exhibits the lowest degree of sequence homology between different receptors and the highest, DBD. Among other domains are those specifying transactivation (AF-1, AF-2), dimerization, nuclear localization signal (NLS) and hsp90 binding site, whose location varies according to whether these functions are ligand-dependent or ligand-independent (Fig. 3.7B). Of particular interest is the property of heterodimerization, especially where it is thought to be functionally important. Several studies have shown that the three isoforms of RXR(α, β and γ) strongly bind to the isoforms of TR, RAR, VDR and PPAR to enhance the interaction of

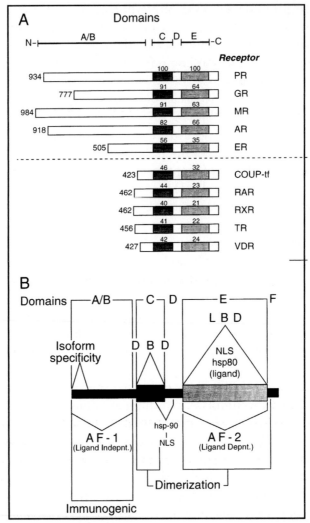

Fig. 3.7. Principal structure-function characteristics of two classes of nuclear steroid/thyroid hormone/retinoid receptors. A. The relative size of the human receptors is indicated by the number of amino acids indicated at the N-terminal end. The modular structure is represented by three major domains: A/B, C, and E. The numbers at top of the bars indicate the % amino acid homology of the consensus regions of the DNA- and ligand-binding domains. B. Some major functions associated with the different domains of nuclear receptors. Much of the immunogenicity of the receptor is localized in the A/B domain, which also contains the ligand-independent transcription activation factor (AF-1). Regions C and D contain the DNA-binding domains, as well as a ligand-independent nuclear localization signal, hsp90 binding and dimerization functions. The same ligand-dependent functions can be detected in domains E and the C-terminal domain F. (See ref. 83 for details.) Abbreviations: LBD, ligand-binding domain; DBD, DNA-binding domain; NLS, nuclear localization signal. Abbreviations for receptors as in Table 3.2.

these receptors with their DNA response elements and augment transcriptional activation.[58,63,72,73,82] Interestingly, RXR will also form homodimers that retain its transactivation response to 9-*cis* retinoic acid. Also of interest is the fact that the natural heterodimeric partner ecdysteroid receptor (EcR) in insects is the product of *ultraspiracle* (*usp*) gene of Drosophila. *usp* is closely related to RXR, such that by domain-swapping the chimeric TR and EcR will now exhibit reciprocal ligand and HRE specificities in insect and mammalian cells.[64-67,73] Besides RXR heterodimers, dimerization among other members of the nuclear receptor gene family has also been reported, such as TR-RAR and VDR-RAR, but their significance in vivo is not yet clear.[72] For further details about the structural and functional aspects of the modular organization of nuclear receptors the reader is referred to several comprehensive reviews and monographs.[3,6,57-59,83-85]

What is it that determines the high degree of target gene specificity associated with a given hormone and its receptor? Many laboratories have intensively studied the DNA sequences with which each receptor, or its DBD, interacts and which is termed hormone-responsive element (HRE). Consensus HRE sequences have now been described for all nuclear receptors, from which several important principles of transcription factor function have emerged. As an extension of the classification of nuclear receptors in Table 3.2, another category can be added, namely that based on the way in which the DBD recognizes its target and how HRE is organized.[63,66,72,86-88] The DBD in all nuclear receptors comprises two highly conserved zinc fingers which constitute a ubiquitous structural motif for nucleic acid recognition.[89] As illustrated in Fig. 3.8A for estrogen receptor[90] of the five conserved cysteines in the 84-amino acid residue DBD, the first four in the second zinc finger act as metal ligands. Amino acid sequences of DBDs of nuclear receptors reveal extended similarities but also what appear to be minor differences. It is the latter which produce different secondary and tertiary folding to allow individual receptors to recognize their target HREs, although the HREs also exhibit both common and divergent features among themselves.[57,58,63,72,90] A consensus hexanucleotide element, usually present as a pair, is the most common feature of HREs, but the sequence of each hexad and the relative position of the two hexads exhibit considerable variability to generate specificity of interaction between the receptor and its target gene. Thus, as shown in Fig. 3.8B, the response elements recognized by steroid hormones (estrogen, progesterone, glucocorticoids, androgen) are characterized by a 15-nucleotide motif consisting of two hexads in a palindromic configuration separated by 3 nucleotides. GRE/PRE differ in two base-pairs of the hexad from that of ERE. However, the HREs of non-steroid receptors of Group II of Table 3.2 (i.e., TR, RAR, RXR, VDR, PPAR) all share the same AGGTCA hexad as ERE but are organized as direct repeats separated by one to five nucleotides.

The arrangement of the HREs of the latter group of receptors is of particular importance, in that it allows a fine discrimination of the target gene by heterodimers with RXRs. This arrangement is the basis for the "1-2-3-4-5 rule". On DR-1 RXR can bind as a homodimer. The partners of RXRs shown in Fig. 3.9 are distinguished by the number of nucleotides separating the two half-sites of the direct repeats. VDR, TR and RAR occupy the 3' half-site, whereas RXR is at 5'.[82,91] Quite interestingly, the ligand-induced transcription by RXR is suppressed when it heterodimerizes with VDR, TR and RAR. The two latter partners actually prevent RXR from binding its ligand 9-*cis*-RA, so that it can be considered as the 'silent' partner.[92]

It is particularly appropriate in the context of the 1-2-3-4-5 rule to take into account the diversity of possible TREs with which TRs are known to interact in different thyroid hormone-regulated genes.[93] Early studies identified a consensus synthetic palindromic sequence AGGTCA.TGACCT, termed TREpal, as the responsive element for TR. However, this was not borne out in later studies which confirmed that most naturally occurring TREs of T_3 target genes have the configuration of tandemly arranged direct repeats (see Fig. 3.10). Various groups have observed that in the absence of the ligand, TR acts as a 'silencer' rather than a positive activator of transcription in the presence of T_3[94,95] (see chapter 4). By contrast, and as also indicated in Fig. 3.10, the genes encoding the thyroid stimulating hormone (TSH) α and β chains and TSH releasing factor (TRF) are negatively regulated in the presence of the ligand.[96-98] This negativity fits in with the physiological process of a negative feedback loop formed when thyroid hormone regulates TSH and TRH expression in the pituitary and hypothalamus, respectively. The rat TSHβ gene has recently been shown to contain distinct response elements for retinoids and thyroid hormone.[125] Negative GREs have also been reported in bovine prolactin and human CRF genes which explain repression of transcription of these genes by glucocorticoid hormone.[99,100] Simi-

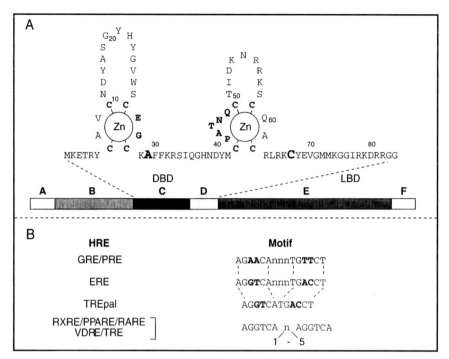

Fig. 3.8. DNA-binding domain (DBD) and the consensus nucleotide sequences of hormone responsive elements (HRE) of some nuclear receptors. A. The amino acid sequence of the DBD of estrogen receptor (single letter amino acid codes), with the conserved cysteines (C) responsible for the Zn-binding sites of the two Zn fingers, written in bold. Also in bold are the amino acids involved in the discrimination of DNA-binding sites. B. A comparison of the consensus nucleotide sequences of the HREs of some nuclear receptors. GRE, PRE and ERE are the palindromic response elements for glucocorticoid, progesterone and estrogen, respectively. The palindromic TREpal is not thought to be a natural TRE. The AGGTCA direct repeats separated by 1, 2, 3, 4 or 5 nucleotides (n) are the natural response elements for the receptors represented at the bottom. The first two letters of the other response elements represent the different receptors and their ligands as represented in Table 3.2. Based on Schwabe JWR, Rhodes D. Beyond zinc fingers: steroid hormone receptors have a novel structural motif for DNA recognition. Trends Biochem Sci 1991; 16:291-96.

larly, the diversity of TREs in different T_3-regulated gene promoters, as opposed to steroid hormone receptors, may be one of the explanations of the multiple physiological actions of thyroid hormones (see chapter 2).

Cell Biology of Nuclear Receptors

The validity of the interaction between DBD of nuclear receptors and their cognate HREs has received dramatic confirmation from NMR (nuclear magnetic resonance) spectroscopy and X-ray crystal structure analysis.[87,101-103] The DBDs of receptors studied so far (ER,

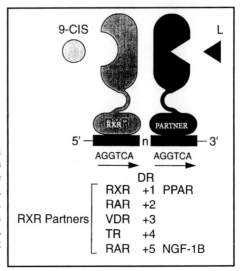

Fig. 3.9. The 1-2-3-4-5 rule. Non-steroidal nuclear receptors that function as heterodimers with RXR to recognize the direct repeat (DR) of the hexad AGGTCA separated by 1-5 nucleotides (n). RXR can form homodimers and NGF-1B is an orphan receptor. L: ligand for RXR partner. See ref. 63 for a detailed account of RXR heterodimers.

GR, TR, RXR) reveal a highly conserved 66-amino acid core comprising the two Zn fingers and a C-terminal extension. This core contains two α helices, one of which makes contact with the bases of the half-site in the major groove of the relevant HRE. Of particular interest is the X-ray crystal structure of the DBDs of TR/RXR heterodimer, resolved by Rastinejad et al[103] (see Fig. 3.11). It clearly highlights a polar head-to-tail arrangement of the DBDs of the two receptors which bind to adjacent major grooves on one side of the DNA double helix. Interestingly, this assembly on direct repeats is in contrast to that shown for the palindromic DNA repeats of response elements of GR and ER which impose a 2-fold symmetry of head-to-head lining up of the DBDs.[87,102]

In addition to heterodimerization of RXR with its principal receptor partners, there are several examples of interactions among other nuclear receptors not involving RXR, as well as between a given receptor and non-receptor transcription factor. Noting that PREs and GREs in the promoters of target genes are often adjacent to binding sites for other transcription factors, such as SP1, NF1 and OTF, Renkawitz's group[104] were able to show that these acted synergistically with liganded PR and GR in inducing transcription. Similarly, both positive and negative regulation of transcriptional activity have been ascribed to co-factors acting synergistically with these and other steroid and retinoid receptors.[105,106] Thus, one has to consider nuclear receptors as part of regulatory complexes formed with their co-factors. Importantly, there are also an increasing number of reports of

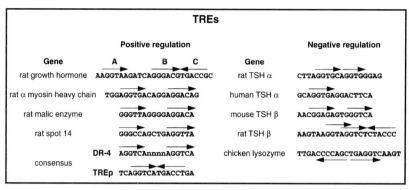

Fig. 3.10. Nucleotide sequences of positive and negative TREs in the promoters of some target genes that are activated or repressed by thyroid hormones, respectively. Abbreviations: TSH, thyroid stimulating hormone gene; DR-4, direct repeat +4 nucleotides; TREp, palindromic responsive element. For details see Chatterjee VKK, Tata JR. Thyroid hormone receptors and their role in development. Cancer Surveys 1992; 14:147-167.

negative regulation of target genes resulting from such interactions among transcription factors, for example, Yamamoto's group's description of GR modulating both positive and negative regulation of the mouse proliferin gene promoter according to its interactions with the factors c-*jun*, c-*fos* and AP-1.[107] Since the activity and nuclear accumulation of transcription factors and nuclear receptors are known to be strongly determined by their phosphorylation status,[108-111] such interactions serve to link up signals transduced through membrane and nuclear receptors. Several laboratories have also recently isolated nuclear factors that act as repressors of nuclear receptors, particularly RAR, RXR and TR, which serve to increase the diversity of tissue-specific and gene-specific responses to different hormonal signals (see chapter 4).

The growing number of complex interactions among regulatory factors focuses attention on the role of chromatin structure in explaining hormone action. It is widely accepted that the higher order of organization of genes within the eukaryotic cell nucleus has an important role to play in transcriptional control.[112] Beato's group have produced evidence that the function of GR in regulating the transcription of MMTV promoter is influenced by the manner in which it is organized in the chromatin structure.[113-115] According to the model in Fig. 3.12, the hormone response elements are highly organized in phased nucleosomes and the binding of glucocorticoids to its nuclear receptor causes an alteration in the chromatin structure, such that it will induce the binding of transcription factors like

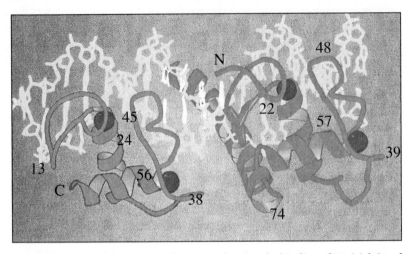

Fig. 3.11. X-ray crystal structure of TR/RXR showing the binding of TR (right) and RXR (left) to the direct repeat +4 (DR+4) sequence of TRE. Also shown are zinc ions (spheres) and the relative disposition of RXR on the 5′ side of response element and the C-terminal projection of C-terminal TR helical tail across the minor groove. Abbreviations: N,C are N- and C-termini; the numbers denote amino acid positions. (From: Rastinejad F, et al. Structural determinants of nuclear receptor assembly on DNA direct repeats. Nature 1995; 375:203-11).

NGF-1 and OTF-1 and thus allow the transcription of MMTV (mouse mammary tumor virus) promoter. A similar, rapidly reversible mechanism of disturbance of nucleosomal organization by the hormone has also been proposed for induction of the rat tyrosine aminotransferase (TAT) gene by glucocorticoids.[116] Both the silencing and activation of transcription of the TRβ gene have also been ascribed to processes controlling nucleosome assembly.[117] According to Beato and his colleagues, this type of chromatin alteration may underlie the transcriptional control by other steroid receptors.[114,115] The conclusions of such studies are largely based on results obtained with crosslinking and determining points of contact between DNA and protein by DNase hypersensitivity assays. While one has to clearly consider the importance of higher order structures to explain transcriptional control, lack of techniques giving more direct information of the spatial organization and mobility of receptors makes it difficult to draw precise conclusions as to the role played by chromatin rearrangements within the nucleus in vivo.

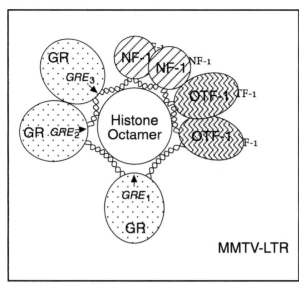

Fig. 3.12. A simplified model of the possible chromatin organization of the MMTV long terminal repeat (LTR), containing multiple GREs and binding sites for other transcription factors (NF-1 and OTF-1). The promoter containing these sites is shown here wrapped around a histone octamer of the nucleosome in a glucocorticoid target cell nucleus. The GR and two transcription factors are shown either to make or not make contact in a highly specific manner. Of the three GR molecules, two are not accessible to the GREs in the absence of the hormone. Truss M, et al. Glucocorticoid hormone induces binding of receptors and transcription factors to a rearranged nucleosome on the MMTV promoter in vivo. EMBO J 1995; 14:1737-51.

Evolution of Receptors

The high degree of conservation of structures and functions of receptors located on the cell surface or intracellularly has attracted the attention of investigators interested in their evolution. Chimeric receptors have been generated for both the membrane and nuclear receptors by a process of domain swapping, which demonstrates that they can regulate each other's functions when linked together. The mosaic structure of receptors strongly suggests that recruitment of exons during evolution must have created new functions for hormones. At the same time, the non-ligand-binding functional domains have been highly conserved during evolution. For example, there is considerable sequence homology between the cytoplasmic domains of the β-subunit of insulin and EGF receptors and the corresponding regions of the *src* oncogene family.[7] There is also evidence that membrane receptors could have co-evolved with their peptide or protein ligands.[118,119] Interestingly, a high degree of sequence homology has been found for 13 or 18 exons of low density lipoprotein receptor (LDLR) with not only receptors of protein hormones (insulin, EGF),

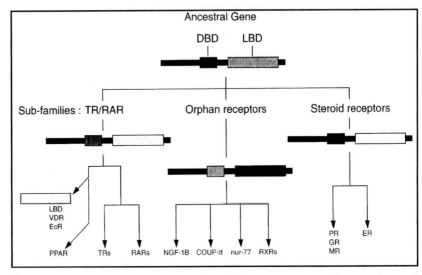

Fig. 3.13. A simplified phylogenetic tree derived from comparisons of DBD and part of LBD of some nuclear receptors, based on the observations of Laudet et al[121] and Amero et al.[122] See text for explanation.

but also with their ligands.[9,120] A characteristic feature is the significant sequence duplication within the ligand-binding domain of the different membrane receptors, which has led to the suggestion that both the protein hormones and their receptors constitute mosaic proteins derived by exon shuffling. The possibility of a common evolutionary origin of membrane receptors involved in growth (EGFR, IR) and nutrient transport into cells (LDLR) is therefore not too far-fetched.

In contrast to the species-dependent variability of membrane receptors, there is a high degree of sequence conservation within the members of nuclear receptor families specified by a given ligand, especially in the DBDs and LBDs. However, it is only a matter of speculation as to whether these two domains evolved independently. One also has to consider the possibility that the early members were primitive, constitutive transcription factors. The fact that RXRs in vertebrates and *usp* in insects are the only nuclear receptors or transcription factors that can heterodimerize with other nuclear receptors suggests that this function must also have evolved early, most likely before vertebrates and invertebrates diverged. To test the idea if nuclear receptor, genes evolved through the duplication of a common ancestral gene, Laudet et al[121] compared phylogenetic trees constructed from the DBD and the part of LBD implicated in transcriptional activation (AF-2) and dimerization functions (see Fig. 3.7B). The tree shown in Fig. 3.13 and based on DBD clearly reveals a common progeny of nuclear receptors which can be grouped into 3 subfamilies: a) TRs/RARs; b) orphan receptors; and c) steroid receptors. The same distribution pattern is evident from a comparison of part of the LBD, except that two groups of receptors are positioned differently than in the DBD-based tree. These include VDR, ECR and NGF-1B genes from which it appears that DBD and LBD belong to different classes. Based on a comparison of DBD only, Amero et al[122] concluded that the nuclear receptors diverged from a common ancestor that contained both DBDs with two homologous Zn fingers and LBDs. They further point out that nuclear receptors do not share a common ancestor with other transcription factors, Zn finger proteins or other ligand binding proteins, such as the serum and cytoplasmic hormone binding proteins. Until more detailed phylogenetic trees are constructed which include additional members of this supergene family, it can be concluded that the evolutionary history of nuclear receptor genes must have involved both gene duplication and domain swapping.

Expression of Receptors

In considering developmental processes, the question of which comes first in ontogeny, i.e., the hormone or its receptor, can be important. This is particularly for postembryonic development. It has been known for some time that a given hormone can induce the expression of the receptor of another which is particularly well-documented for steroid hormones (see ref. 123), which can explain the synergestic or additive actions of multiple hormones regulating the growth and development of the same tissue. There are also some reports of negative cross-regulation of hormone receptors. More interesting is the phenomenon of autoregulation of a receptor by its own ligand which can be of considerable physiological significance. These questions of receptor gene expression constitute the topic of two later chapters (see chapters 6 and 7).

References

1. Schulster D, Levitzki A, eds. Cellular Receptors. Chichester: John Wiley, 1980.
2. Norman AW, Litwack G. Hormones. Orlando, Academic Press, 1987.
3. Baulieu E-E, Kelly P, eds. Hormones. From Molecules to Disease. Paris: Hermann, 1990.

4. O'Malley BW, Birnbaumer L, eds. Receptors and Hormone Action. New York: Academic Press, 1978:v1-3.

5. Rickenberg HV, ed. International Review of Biochemistry, Biochemistry and Model of action of Hormones II, Baltimore: University Park Press, 1978:v20.

6. Parker MG, ed. Nuclear Hormone Receptors. London: Academic Press, 1991.

7. Ullrich A, Bell JR, Chen EY et al. Human insulin receptor and its relationship to the tyrosine kinase family of oncogenes. Nature 1985; 313:756-61.

8. Downward J, Yarden Y, Mayes E et al. Close similarity of epidermal growth factor receptor and v-erbB oncogene protein sequences. Nature 1984: 307:521-27.

9. Ebina Y, Ellis L, Jarnagin K et al. The human insulin receptor cDNA: The structural basis for hormone-activated transmembrane signaling. Cell 1985; 40:747-58.

10. Hsuan JJ, Panayotou G, Waterfield MD. Structural basis for epidermal growth factor receptor function. Progr Growth Factor Res 1989; 1:23-32.

11. Kahn CR, White MF, Shoelson SE et al. The insulin receptor and its substrate: molecular determinants of early events in insulin action. Rec Progr Horm Res 1993; 48:291-339.

12. Kelly PA, Ali S, Rozakis M. The growth hormone/prolactin receptor family. Rec Progr Horm Res 1993; 48:123-64.

13. Parker P, ed. Cell signaling. Cold Spring Harbor: Cold Spring Harbor Laboratory Press, 1996.

14. Hunter T. Oncoprotein networks. Cell 1997; 88:333-46.

15. van der Geer P, Hunter P, Lindberg RA. Receptor protein tyrosine kinases and their signal transduction pathways. Ann Rev Cell Biol 1994; 10:251-337.

16. Sutherland EW. Studies on the mechanism of hormone action. Science 1972; 177:401-08.

17. Berridge MJ. Inositol lipids and calcium signaling. Proc R Soc Lond B 1988; 234:359-78.

18. Caron MG, Lefkowitz RJ. Catecholamine receptors: Structure, function and regulation. Rec Progr Horm Res 1993; 48:277-90.

19. Cohen P, Foulkes JG, eds. The Hormonal Control of Gene Transcription. Amsterdam: Elsevier, 1992.

20. Barritt GJ. Communication Within Animal Cells. Oxford: Oxford University Press, 1992.

21. Stamler JS, Singel DJ, Loscalzo J. Biochemistry of nitric oxide and its redox-activated forms. Science 1992; 258:1898-902.

22. Irvine RF, Michell RH, Marshall CJ. Current understanding of intracellular signaling pathways. Phil Trans R Soc Lond B 1996; 351:123-41.

23. Moncada S, Stamler J, Gross S et al, eds. The Biology of Nitric Oxide: Part 5. London: Portland Press, 1996.

24. Briscoe J, Guschin D, Rogers NC et al. JAKs, STATs and signal transduction in response to the interferons and other cytokines. Phil Trans R Soc Lond B 1996; 351:167-71.
25. Clevenger CV, Medaglia MV. The protein tyrosine kinase p59[fyn] is associated with prolactin (PRL) receptor and is activated by PRL stimulation of T-lymphocytes. Mol Endocrinol 1994; 8:674-81.
26. Campbell JS, Seger R, Graves JD et al. The MAP kinase cascade. Rec Progr Horm Res 1995; 50:131-59.
27. Errede B, Ge Q-Y. Feedback regulation of map kinase signal pathways. Phil Trans R Soc Lond B 1996; 351:143-49.
28. White MF. The IRS-signaling system in insulin and cytokine action. Phil Trans R Soc Lond B 1996; 351:181-89.
29. Quon MJ, Butte AJ, Taylor SI. Insulin signal-transduction pathways. Trends Endocrinol Met 1994; 5:369-76.
30. Welte T, Garimorth K, Philipp S et al. Prolactin-dependent activation of a tyrosine phosphorylated DNA binding factor in mouse mammary epithelial cells. Mol Endocrinol 1994; 8:1091-1102.
31. Ihle JN. Signaling by the cytokine receptor superfamily—just another kinase story. Trends Endocrinol Met 1994; 5:137-43.
32. Daly C, Reich NC. Receptor to nucleus signaling via tyrosine phosphorylation of the P91 transcription factor. Trends Endocrinol Met 1994; 5:159-64.
33. Montminy MR, Gonzalez GA, Yamamoto KK. Characteristics of the cAMP response unit. In: Cohen P, Foulkes JG, eds. The Hormonal Control of Gene Transcription. Amsterdam: Elsevier, 1991:161-71.
34. Karin M. The regulation of AP-1 activity by mitogen-activated protein kinases. Phil Trans R Soc Lond B 1996; 351:12734.
35. Collins S, Caron MG, Lefkowitz RJ. From ligand binding to gene expression: new insights into the regulation of G-protein-coupled receptors. Trends Biochem Sci 1992; 17:37-39.
36. Jensen EV, Jacobson HI. Basic guides to the mechanism of estrogen action. Rec Progr Horm Res 1962; 18:387-414.
37. Talwar GP, Segal SJ, Evans A et al. The binding of estradiol in the uterus: a mechanism for derepression of RNA synthesis. Proc Natl Acad Sci USA 1964; 52:1059-66.
38. Toft D, Gorski J. A receptor molecule for estrogens: Isolation from the rat uterus and preliminary characterization. Proc Natl Acad Sci USA 1966; 55:1574-81.
39. De Sombre ER, Puca GA, Jensen EV. Purification of an estrophilic protein from calf uterus. Proc Natl Acad Sci USA 1969; 64:148-54.
40. Jensen EV, Suzuki T, Kawashima T et al. A two-step mechanism for the interaction of estradiol with rat uterus. Proc Natl Acad Sci USA 1968; 59:632-38.
41. Baulieu E-E, Atger M, Best-Belpomme M et al. Steroid hormone receptors. Vitamins and Hormones. New York: Academic Press, 1975; 649-736.

42. Milgrom E, Thi L, Atger M et al. Mechanisms regulating the concentration and the conformation of progesterone receptor(s) in the uterus. J Biol Chem 1973; 248:6366-74.
43. Raaka BM, Samuels HH. The glucocorticoid receptor in GH₁ cells. J Biol Chem 1983; 258:417-25.
44. Sar M, Liao S, Stumpf WE. Nuclear concentration of androgens in rat seminal vesicles and prostate demonstrated by dry-mount autoradiography. Endocrinology 1970; 86:1008-11.
45. King WJ, Greene GL. Monoclonal antibodies localize oestrogen receptor in the nuclei of target cells. Nature 1984; 307:745-47.
46. Sanchez ER, Toft DO, Schlesinger MJ et al. Evidence that the 90-kDa phosphoprotein associated with the untransformed L-cell glucocorticoid receptor is a murine heat shock protein. J Biol Chem 1985; 260:12398-401.
47. Baulieu E-E. Steroid hormone antagonists at the receptor level: A role for the heat-shock protein MW 90,000 (hsp 90). J Cell Biochem 1987; 35:161-74.
48. Guichon-Mantel A, Delabre K, Lescop P et al. Intracellular traffic of steroid hormone receptors. J Ster Biochem Mol Biol 1996; 56:3-9.
49. Hollenberg SM, Weinberger C, Ong ES et al. Primary structure and expression of a functional human glucocorticoid receptor cDNA. Nature 1985; 318:635-41.
50. Green S, Walter P, Kumar V et al. Human oestrogen receptor cDNA: sequence, expression and homology to v-erbA. Nature 1986; 320:134-39.
51. Danielsen M, Northrop JP, Ringold GM. The mouse glucocorticoid receptor: mapping of functional domains by cloning, sequencing and expression of wild-type and mutant receptor proteins. EMBO J 1986; 5:2513-22.
52. Yamamoto KR. Steroid receptor regulated transcription of specific genes and gene networks. Annu Rev Genet 1985; 19:209-52.
53. Evans RM. The steroid and thyroid hormone receptor superfamily. Science 1988; 240:889-95.
54. Sap J, Munoz A, Damm K et al. The c-erb-A protein is a high-affinity receptor for thyroid hormone. Nature:London 1986; 324:635-640.
55. Weinberger C, Thompson CC, Ong ES et al. The c-erb-A gene encodes a thyroid hormone receptor. Nature 1986; 324: 641-46.
56. Green S, Chambon P. Nuclear receptors enhance our understanding of transcription regulation. Trends Genet 1988; 4:309-14.
57. Carson-Jurica MA, Schrader WT, O'Malley BW. Steroid receptor family: structure and functions. Endocrine Rev 1990; 11:201-20.
58. Mangelsdorf DJ, Thummel C, Beato M et al. The nuclear receptor superfamily: The second decade. Cell 1995; 83:835-39.
59. Ribeiro RCJ, Kushner PJ, Baxter JD. The nuclear hormone receptor gene superfamily. Annu Rev Med 1995; 46:443-53.
60. Conneely OM, O'Malley BW. Orphan receptors: structure and function relationships. In: Tsai M-J, O'Malley BW, eds. Mechanism of

Steroid Hormone Regulation of Gene Transcription. Austin: RG Landes, 1994:111-33.

61. Heyman RA, Mangelsdorf DJ, Dyck JA et al. 9-cis retinoic acid is a high affinity ligand for the retinoid X receptor. Cell 1992; 68:397-406.

62. Zhang X, Hoffmann B, Tran PH-V et al. Retinoid X receptor is an auxiliary protein for thyroid hormone and retinoic acid receptors. Nature 1992; 355:441-46.

63. Mangelsdorf DJ, Evans RM. The RXR heterodimers and orphan receptors. Cell 1995; 83:841-50.

64. Oro AE, McKeown M, Evans RM. Relationship between the product of the Drosophila ultraspiracle locus and vertebrate retinoid X receptor. Nature 1990; 347:298-301.

65. Thomas HE, Stunnenberg HG, Stewart AF. Heterodimerization of the Drosophila ecdysone receptor with retinoid X receptor and ultraspiracle. Nature 1993; 362:471-75.

66. Segraves WA. Steroid receptors and other transcription factors in ecdysone response. Rec Progr Horm Res 1994; 49:167-95.

67. Thummel CS. From embryogenesis to metamorphosis: The regulation and function of Drosophila nuclear receptor superfamily members. Cell 1995; 83:871-77.

68. Qiu Y, Tsai SY, Tsai M-J. COUP-TF. An orphan member of the steroid/thyroid hormone receptor superfamily. Trends in Endocrin 1994; 5:234-39.

69. Issemann I, Green S. Activation of a member of the steroid hormone receptor superfamily by peroxisome proliferators. Nature 1990; 347:645-50.

70. Dreyer C, Krey G, Keller H et al. Control of the peroxisomal -oxidation pathway by a novel family of nuclear hormone receptors. Cell 1992; 68:879-87.

71. Fawell SE, Lees JA, White R et al. Characterization and colocalization of steroid binding and dimerization activities in the mouse estrogen receptor. Cell 1990; 60:953-62.

72. Glass CK. Differential recognition of target genes by nuclear receptor monomers, dimers and heterodimers. Endocrin Rev 1994; 15:391-407.

73. Yu VC, Delsert C, Andersen B et al. RXRβ: A coregulator that enhances binding of retinoic acid, thyroid hormone, and vitamin D receptors to their cognate response elements. Cell 1991; 67:1251-66.

74. Kastner P, Mark M, Chambon P. Nonsteroid nuclear receptors: What are genetic studies telling us about their role in real life? Cell 1995; 83:859-69.

75. Leid M, Kastner P, Chambon P. Multiplicity generates diversity in the retinoic acid signaling pathways. Trends Biochem Sci 1992; 17:427-33.

76. Lufkin T, Lohnes D, Mark M et al. High postnatal lethality and testis degeneration in retinoic acid receptor α (RARα) mutant mice. Proc Natl Acad Sci USA 1993; 90:7225-29.

77. Lee SS-T, Pineau T, Drago J et al. Targeted disruption of the α isoform of the peroxisome proliferator-activated receptor gene in mice results in abolishment of the pleiotropic effects of peroxisome proliferators. Mol Cell Biol 1995; 15:3012-22.

78. Brun RP, Tontonoz P, Forman BM et al. Differential activation of adipogenesis by multiple PPAR isoforms. Genes Dev 1996; 10:974-84.

79. Forrest D, Hanebuth E, Smeyne RJ et al. Recessive resistance to thyroid hormone in mice lacking thyroid hormone receptor β: Evidence for tissue-specific modulation of receptor function. EMBO J 1996; 15:3006-15.

80. Lohnes D, Mark M, Mendelsohn C et al. Function of the retinoic acid receptors (RARs) during development. I. Craniofacial and skeletal abnormalities in RAR double mutants. Development 1994; 120:2723-48.

81. Kumar V, Green S, Stack G et al. Functional domains of the human estrogen receptor. Cell 1987; 51:941-51.

82. Zechel C, Shen X-Q, Chen J-Y et al. The dimerization interfaces formed between the DNA binding domains of RXR, RAR and TR determine the binding specificity and polarity of the full-length receptors to direct repeats. EMBO J 1994; 13:1425-33.

83. Chambon P. The molecular and genetic dissection of the retinoid signaling pathway. Rec Progr Horm Res 1995; 50:317-32.

84. Gronemeyer H. Control of transcription activation by steroid hormone receptors. FASEB J 1992; 6:2524-29.

85. O'Malley BW, Tsai SY, Bagchi M et al. Molecular mechanism of action of a steroid hormone receptor. Rec Progr Horm Res 1991; 47:1-24.

86. Laudet V, Stehelin D. Nuclear receptors. Flexible friends. Current Biology 1992; 2:293-95.

87. Schwabe JWR, Chapman L, Finch JT et al. The crystal structure of the estrogen receptor DNA-binding domain bound to DNA: How receptors discriminate between their response elements. Cell 1993; 75:567-78.

88. Strahle U, Klock G, Schutz G. A DNA sequence of 15 base pairs is sufficient to mediate both glucocorticoid and progesterone induction of gene expression. Proc Natl Acad Sci USA 1987; 84:7871-75.

89. Klug A, Schwabe JWR. Zinc fingers. FASEB J 1995; 9:597-604.

90. Schwabe JWR, Rhodes D. Beyond zinc fingers: Steroid hormone receptors have a novel structural motif for DNA recognition. Trends Biochem Sci 1991; 16:291-96.

91. Kurokawa R, Yu VC, Naar A et al. Differential orientations of the DNA-binding domain and carboxy-terminal dimerization interface regulate binding site selection by nuclear receptor heterodimers. Genes Dev 1993; 7:1423-35.

92. Leblanc BP, Stunnenberg HG. 9-cis retinoic acid signaling: Changing partners causes some excitement. Genes Dev 1995; 9:1811-16.

93. Chatterjee VKK, Tata JR. Thyroid hormone receptors and their role in development. Cancer Surveys 1992; 14:147-167.

94. Baniahmad A, Tsai SY, O'Malley BW et al. Kindred S thyroid hormone receptor is an active and constitutive silencer and a repressor for thyroid hormone and retinoic acid responses. Proc Natl Acad Sci USA 1992; 89:10633-7.

95. Fondell JD, Roy AL, Roeder RG. Unliganded thyroid hormone receptor inhibits formation of a functional preinitiation complex: implications for active repression. Genes Dev 1993; 7:1400-10.

96. Chatterjee VKK, Lee J-K, Rentoumis A, Jameson JL. Negative regulation of the thyroid stimulating hormone a gene by thyroid hormone: receptor interaction adjacent to the TATA box. Proc Natl Acad Sci USA 1989; 86:9114-18.

97. Hollenberg AN, Monden T, Flynn TR et al. The human thyrotropin-releasing hormone gene is regulated by thyroid hormone through two distinct classes of negative thyroid hormone response elements. Mol Endocrinol 1995; 9:540-50.

98. Safer JD, Langlois MF, Cohen R. Isoform variable action among thyroid hormone receptor mutants provides insight into pituitary resistance to thyroid hormone. Mol Endocrinol 1997; 11:16-26.

99. Sakai DD, Helms S, Carlstedt-Duke J et al. Hormone-mediated repression: a negative glucocorticoid response element from the bovine prolactin gene. Genes Dev 1988; 2:1144-54.

100. Malcoski SP, Handanos CPM, Dorin RJ. Localization of a negative glucocorticoid response element of the human corticotropin releasing hormone gene. Mol Cell Endocrinol 1997; 127:189-99.

101. Lee MS, Kliewer SA, Provencal J et al. Structure of the retinoid X receptor a DNA binding domain: A helix required for homodimeric DNA binding. Science 1993; 260:1117-21.

102. Luisi BF, Xu WX, Otwinowski Z et al. Crystallographic analysis of the interaction of the glucocorticoid receptor with DNA. Nature 1991; 352:497-505.

103. Rastinejad F, Perlmann T, Evans RM et al. Structural determinants of nuclear receptor assembly on DNA direct repeats. Nature 1995; 375:203-11.

104. Schule R, Muller M, Kaltschmidt C et al. Many transcription factors interact synergistically with steroid receptors. Science 1988; 242: 1418-20.

105. Herrlich P, Ponta H. Mutual cross-modulation of steroid/retinoic acid receptor and AP-1 transcription factor activities. Trends Endocrinol Met 1994; 5:341-46.

106. Folkers GE, van den Burg, van den Saag. A role for cofactors in synergistic and cell-specific activation by retinoic acid receptors and retinoid X receptor. J Ster Biochem Mol Biol 1996; 56:119-29.

107. Diamond MI, Miner JN, Yoshinaga SK et al. Transcription factor interactions: selectors of positive or negative regulation from a single DNA element. Science 1990; 249:1266-72.

108. Auricchio F. Phosphorylation of steroid receptors. J Ster Biochem 1989; 32:613-22.

109. Goldberg Y, Glineur C, Gesquiere J-C et al. Activation of protein kinase C or cAMP-dependent protein kinase increases phosphorylation of the c-erbA-encoded thyroid hormone receptor and of the v-erbA-encoded protein. EMBO J 1988; 7:2425-33.

110. Aronica SM, Katzenellenbogen BS. Stimulation of estrogen receptor-mediated transcription and alteration in the phosphorylation state of the rat uterine estrogen receptor by estrogen, cyclic adenosine monophosphate, and insulin-like growth factor-1. Mol Endocrin 1993; 7:743-52.

111. Arnold SF, Melamed M, Vorojeikina DP et al. Estradiol-binding mechanism and binding capacity of the human estrogen receptor is regulated by tyrosine phosphorylation. Mol Endocrinol 1997; 11:48-53.

112. Wolffe AP. Regulation of Chromatin Structure and Function. Austin: RG Landes, 1994.

113. Beato M, Herrlich P, Schutz G. Steroid hormone receptors: many actors in search of a plot. Cell 1995; 83:851-57.

114. Beato M, Candau R, Chavez S et al. Interaction of steroid hormone receptors with transcription factors involves chromatin remodelling. J Ster Biochem MolBiol 1996; 56:47-59.

115. Truss M, Bartsch J, Schelbert A et al. Glucocorticoid hormone induces binding of receptors and transcription factors to a rearranged nucleosome on the MMTV promoter in vivo. EMBO J 1995; 14:1737-51.

116. Reik A, Schutz G, Stewart AF. Glucocorticoids are required for establishment and maintenance of an alteration in chromatin structure: induction leads to a reversible disruption of nucleosomes over an enhancer. EMBO J 1991; 10:2569-76.

117. Wong J, Shi Y-B, Wolffe AP. A role for nucleosome assembly in both silencing and activation of the *Xenopus* TRβA gene by the thyroid hormone receptor. Genes Dev 1995; 9:2696-2711.

118. Blundell TL, Humbel RE. Hormone families. Pancreatic hormones and homologous growth factors. Nature (London) 1980; 287:781-87.

119. Tata JR. Evolution of hormones and their actions. Chem Scripta 1986; 26B:179-90.

120. Sudhof TC, Goldstein JL, Brown MS et al. The LDL receptor gene. A Mosaic of exons shaved with other proteins. Science 1985; 228:815-22.

121. Laudet V, Hanni C, Coll J et al. Evolution of the nuclear receptor gene superfamily. EMBO J 1992; 11:1003-13.

122. Amero SA, Kretsinger RH, Moncrief ND et al. The origin of nuclear receptor proteins: a single precursor distinct from other transcription factors. Mol Endocrinol 1992; 6:3-7.

123. Tata JR. The action of growth and developmental hormones. Biol Rev 1980; 55:285-331.

124. Enmark E, Pelto-Huikko M, Nilsson S et al. Cloning of a novel estrogen receptor expressed in rat prostate and ovary. Proc Natl Acad Sci USA 1996; 93:5925-5930.

125. Breen JJ, Hickok NJ, Gurr JA. The rat TSHβ gene contains distinct response elements for regulation by retinoids and thyroid hormone. Mol Cell Endocrin 1997; 131:137-146.

CHAPTER 4

Hormonal Regulation of Transcription

The enormous interest over the last 60 years in the subject of hormone action has led to the emergence of some fundamental principles of cellular regulation. To cite only a few examples: the discoveries of cyclic AMP, Ca^{2+}, inositol trisphosphate (IP_3) and nitric oxide as second messengers have largely resulted from attempts to explain how certain hormones exert their physiological actions. Our present-day understanding of the importance of water transport and ions into and out of cells in response to extracellular stimuli also owes its origin to studies on hormonal control of metabolic activity. Similarly, interest in explaining the growth and developmental actions of hormones in the wake of the advent of techniques of molecular biology has been responsible for many current concepts of control of eukaryotic RNA and protein synthesis. In the context of hormonal regulation of developmental processes, the response of the transcriptional machinery to hormonal signaling is the single most important mechanism. However, before considering the hormonal regulation of transcription in detail, it is worth considering briefly the historical evolution of our thinking on hormone action.

A Brief Historical Background

Early ideas of unique mechanisms to account for the actions of all hormones were based on complex cellular processes, such as glycolysis, respiration, etc. The increasing availability of pure enzymes in the 1930s and 1940s saw a shift towards individual biochemical reactions based on single hormone-enzyme interactions.[1] Studies on these interactions led to the proposal that most hormones act via a common mechanism of inducing allosteric or conformational changes in enzymes to regulate cellular activity.[2] A major experimental drawback of work on direct interactions between hormones

Hormonal Signaling and Postembryonic Development,
by Jamshed R. Tata. © 1998 Springer-Verlag and R.G. Landes Company.

(or other signaling molecules) and purified enzymes was the very high concentrations of hormones needed to elicit an effect in vitro which was highly incompatible with the physiological actions of hormones. By the early 1960s there was little enthusiasm for direct hormone-enzyme interaction studies (see Fig. 4.1). Meanwhile in the early 1940s, Levine (see ref. 3) had proposed that insulin was more likely to control sugar metabolism by virtue of its ability to regulate glucose transport into the cell, than by directly interacting in the cytoplasm with hexokinase, then a popular target for hormone-enzyme interactions.[1] Since then, it has been recognized that altered permeability to sugars, amino acids, ions and water is one of the early pleiotypic responses of target cells to hormones.[4] This recognition is the forerunner to our current ideas on the cell membrane being a prime site of hormone action (see chapter 3).

Undoubtedly, it is the discovery of cyclic AMP in Sutherland's laboratory in the mid-1950s, in the course of their studies on how epinephrine and glucagon modulated glycolytic activity of liver and other tissue has had the most profound conceptual impact in our thinking about hormone action.[5] The ubiquitous presence of adenylyl cyclase in eukaryotic and prokaryotic cell membranes is perhaps the most valuable direct spin-off of research on mechanisms of hormone action to the biochemistry of cellular regulation. Cyclic AMP is a key second messenger facilitating a rapid response to changes in a cell's environment, including both chemical and physical signals, such as nutrient levels, light, temperature, pheromones, etc. By its direct involvement in these rapid responses, it has also highlighted the importance of protein phosphorylation in virtually every major cellular and neural function.[6] The concept of second messengers has since been extended to IP_3, Ca^{2+}, G-protein and nitric oxide.[7,8]

In the excitement surrounding the discovery of cyclic AMP, another key observation concerning hormone action remained relatively ignored. In 1956, Knox concluded that glucocorticoid hormone regulated hepatic tyrosine aminotransferase (TAT) by specifically promoting the synthesis of the enzyme.[9] The importance of this observation was only fully realized later when Tomkins exploited rat hepatoma cells in culture to characterize the induction of TAT by the synthetic glucocorticoid dexamethasone (Dex) based on a generalized model of hormonal regulation of protein synthesis was proposed.[10] It was however the establishment of cell-free translation systems that allowed a more precise analysis of how growth and de-

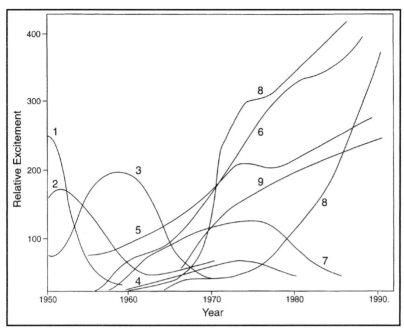

Fig. 4.1. Schematic representation of how our concepts of mechanisms of hormone action have evolved during the 40 years since 1950. "Relative excitement" is based on frequency of publications, topics of specialized conferences, etc. Concepts: 1, Cellular respiration; 2, hormone-enzyme interactions; 3, oxidative phosphorylation and mitochondrial respiration; 4, allosteric mechanisms; 5, changes in membrane structure and function; 6, cyclic AMP and other second messengers; 7, translational control; 8, transcriptional control and (later) transcription factors; 9, gene and chromatin structure. Based on ref. 3. Tata JR. The search for the mechanism of hormone action. Pers Biol Med 1986; 29:184-204.

velopmental hormones influenced protein synthesis in their target cells. In this context, worth citing are the pioneering studies of Korner on the stimulation of amino acid incorporation into proteins by ribosomal preparations from tissues of hypophysectomized rats following treatment with growth hormone.[11] Several investigators were later able to demonstrate a similar response of target tissues of other growth-promoting and developmental hormones such as estrogen, androgen, thyroid hormone and ecdysteroids.[12] Although a direct interaction between a given hormone and the target cell's translational machinery had been proposed in earlier studies, this notion was discarded by the majority of investigators by 1970 (see Fig. 4.1). The reason for this change in thinking stems from a major characteristic of the findings on protein synthesis, namely that the stimulation of amino acid incorporation in vitro following hormonal treatment in vivo could not be reproduced by adding physiologically meaningful amounts of hormones directly to the relevant cell-free system (see Table 4.1). At the same time it became increasingly evident that hormonal treatment influenced the composition of polyribosomes both qualitatively and quantitatively.[13]

With the availability of inhibitors of RNA and protein synthesis (actinomycin D, α-amanitin, puromycin, cycloheximide), it was possible to show that much of the effect of hormones on protein synthesis could be explained on the basis of transcriptional control.[13-16] For example, all physiological actions of thyroid hormones, not only growth and maturation of mammals and amphibian metamorphosis, but also the regulation of oxidative metabolism and basal metabolic rate could be substantially blocked by actinomycin D (Table 4.2 and see refs. 13,17). Similar inhibition of growth or induction of specific enzymes or specialized proteins in their target tissues by estrogen, androgen, glucocorticoids and ecdysteroids were soon demonstrated.[13,15,16,18,19] The massive build-up of polyribosomes and t-RNA in target tissues that takes place following hormone administration was also shown to be blocked by actinomycin D in vivo and in tissue slices. Indirectly, kinetics of labeling of nuclear RNA and polyribosomes revealed that growth and developmental hormones strongly influence the formation and turnover of messenger RNA.

Hormonal Control of Overall Transcriptional Activity

The first indication that hormones regulate transcription came from gene puffing in polytene chromosomes of some insects. In the salivary glands of *Drosophila* and *Chironomus* metamorphosis is characterized by a series of highly specific and sequential chromosomal puffs which are under the control of the molting hormone ecdysone (see chapter 5). These puffs, which contain newly synthesized RNA, can be precociously induced by ecdysteroids and there is now much literature to support the mechanism by which these insect hormones control both the overall rate of RNA synthesis and activate a specific sets of genes.[20-23]

The failure of most hormones to modify cytoplasmic protein synthesis when added directly to cell-free preparations and the strong inhibition by actinomycin D of tissue responses to hormones administered in vivo (Table 4.2), led many investigators in the 1960s to

Table 4.1. *Examples of stimulation of overall protein synthesis and induction of specific enzymes or proteins in target tissues by growth and developmental hormones administered in vivo but which could not be reproduced when the hormone was added to tissue preparations in vitro*

Hormone	Target Tissue	Stimulation of Overall Protein Synthesis in Tissue Preparations	Induction of Specific Enzymes or Proteins In Vivo
Corticosteroid	Rat liver	Slices, microsomes	Tyrosine aminotransferase; tryptophan pyrrolase
Thyroid hormone	Rat liver, muscle, kidney, heart	Slices, polysomes, microsomes	Mitochondrial respiratory enzymes
	Frog tadpole liver	Slices, polysomes	Urea cycles enzymes
	Frog tadpole tail	Organ explants	Proteases, nucleases
Estrogen	Rat, mouse uterus	Slices, homogenates, polysomes	Overall protein synthesis
	Chicken liver, oviduct	Slices, polysomes	Vitellogenin, ovalbumin, conalbumin
	Frog liver, oviduct	Polysomes	Vitellogenin, FOSP-1
Ecdysone	Blowfly, *Drosophila* fat body, epidermis, ovary	Homogenates, polysomes	Egg proteins, cuticle proteins; DOPA decarboxylase
Testosterone	Rat, mouse prostate, seminal vesicles	Slices, homogenates, polysomes	Prostate specific antigen, steroid-binding protein
	Mouse liver	Slices, polysomes	Urinary a_{2u}-globulin

Table 4.2. Early work with actinomycin D demonstrating that RNA synthesis is important for the physiological action of a variety of growth and developmental hormones

Hormone	Tissue	Inhibition of Hormonal Stimulation of:
ACTH	Rat adrenal	Corticosterone production
TSH	Sheep thyroid slices	Thyroid hormone synthesis
Thyroid hormone	Rat liver, muscle, etc.	BMR, growth, cellular respiration
	Frog tadpole liver, tail organ cultures	All metamorphic responses, tail regression
Testosterone	Rat muscle, male accessory tissues	Growth
Estrogen	Rat uterus	Growth, lipid synthesis
	Chicken liver	Egg protein induction
Glucocorticoids	Rodent liver	Induction of TAT, tryptohan pyrrolase, gluconeogenesis
Parathyroid hormone	Rat bone	Bone Ca and P metabolism
Aldosterone	Toad bladder, skin	Na^+ transport
	Rat kidney	Antidiuretic effect
Gonadotropin	Rat testis	Growth of immature testis
Ecdysone	Insect larval tissues	Pupation, chromosomal puffing, all metamorphic responses
Erythropoietin	Human marrow cell cultures	Hemoglobin synthesis DNA synthesis

In all experiments, the hormone and actinomycin D were administered together in vivo or in organ and tissue culture. Based on work carried out prior to 1966.

focus on the possible hormonal control of total RNA synthesis.[12–14] Another reason for investigating transcription was the relatively long latent period that elapsed between the administration of hormone and the manifestation of enhanced protein synthetic activity or the induction of specific enzymes in the target tissues. Prior to the widespread availability of recombinant DNA technology, investigations on transcription were largely restricted to the measurements of accumulation of newly synthesized RNA in the cytoplasm, the kinetics of synthesis and turnover of rapidly labeled nuclear RNA and the accumulation of labeled RNA in the polysomal and tRNA fractions. At the same time, techniques for measuring RNA synthesis by isolated nuclei or run-on/run-off transcription were being developed, which provided an additional tool for analyzing the transcriptional responses to hormones.

Table 4.3. Results of early studies on hormonal stimulation of rapidly-labeled nuclear RNA (R-L RNA) synthesized in vivo and as measured as run-on transcription[1]

Hormone[2]	Target tissue	Lag period[3]		Max. Stimulation %	
		R-L RNA	Run-on	R-L RNA	Run-on
Estrogen	Rat uterus	10 min	1 h	500	150
	Chick oviduct	-	24 h	-	1100
Hydrocortisone	Rat liver	1 h	3 h	300	75
Testosterone	Rat prostate	1.5 h	2.5 h	400	20
Thyroid hormone	Rat liver	3 h	8 h	350	200
	Bullfrog tadpole liver	8 h	-	600	-

[1] For details see refs. 13, 15, 92–95.
[2] Hormone was administered to immature or hormone-deprived animals.
[3] Lag period is the time elapsed between hormone administration and a 10% increase in the rate of RNA synthesis.

In data gathered in the 1960s and summarized in Table 4.3, virtually every growth and developmental hormone administered enhanced the rate of RNA synthesized in vivo in tissue slices, as well as by measuring run-off or run-on transcription in nuclei isolated from target tissues. Although the rapidly labeled nuclear RNA (RL-RNA) fraction would represent all species of newly synthesized RNA, in most of the studies, the labeling period was very short (2-10 min) so that mRNA would represent a major component of this fraction. This was verified in early studies by hybridization with sub-fractions of polysomal RNA (it was not possible to carry out direct determination of mRNA in the 1960s; below). As regards run-on assays, the enhancement of overall transcriptional activity of isolated nuclei from animals or cells treated with a given hormone was later manifested and with a lower stimulation than when assessed by labeling nuclear RNA in vivo or in tissue slices. Not shown in Table 4.3 is that in many of the examples illustrating the in vitro nuclear transcription determinations giving quantitatively different results according to the ionic conditions used. Using the technique of nearest neighbor frequency analysis, Widnell and Tata[24] had shown in 1965 that the major RNA product formed by incubation of nuclear preparations at low ionic strength with Mg^{2+} is rRNA, whereas at high ionic strength with Mn^{2+} it was mRNA (then termed DNA-like RNA). Indeed it was during an investigation of the effect of thyroid hormone RNA synthesis by isolated nuclei that these workers were able to first show the multiplicity of eukaryotic RNA polymerase.[25] By exploiting ion-exchange chromatography and differential inhibition of RNA polymerases by α-amanitin, other workers were later able to definitively establish the identities of RNA polymerases I, II and III.[26,27]

The lag or latent periods shown in Table 4.3 is of special significance because the overall transcriptional stimulations were the earliest biochemical responses to these (and other) growth and developmental hormones, being manifested substantially before other biochemical responses such as protein synthesis, oxidative phosphorylation, glycolysis, membrane proliferation, etc. Until it was shown that actinomycin D blocked the main physiological action of thyroid hormones in adult rats, namely control of basal metabolic rate (Table 4.2), the mechanism of action of TH was widely accepted as a direct action on mitochondrial respiration.[17,28] It is therefore useful to consider the sequence of biochemical responses leading to the enhanced respiratory activity of mitochondria provoked by a single injection of thyroid hormone to young thyroidectomized rats (Fig. 4.2). The enhanced rate of transcription precedes by several hours the appearance of newly synthesized RNA in the cytoplasm, enhanced accumulation of polyribosomes, proliferation of intracellular membranes (mitochondria, microsomes), stimulation of protein synthesis and oxidative phosphorylation (leading to increased tissue respiration), in that order.[17,29] A similar sequence of responses of target tissues involving an increased rate of protein synthesis and growth or the induction of specific proteins was observed for other growth-promoting hormones particularly steroid hormones (see refs. 12,30-32, and Table 4.1). Thus the early studies on total RNA synthesis strongly indicated that the primary action of growth and developmental hormones somehow involved the regulation of transcriptional machinery.

Hormonal Regulation of Specific Gene Expression

Recording the stimulation of the rate of total RNA synthesis is of limited value in exploring the mechanism by which hormones regulate the activity of specific genes. One had to await the advent of gene cloning and recombinant DNA technology to undertake such investigations. It is undoubtedly the chicken estrogen-ovalbumin system that laid the foundation of our current knowledge of hormonal regulation of tissue-specific expression of well-defined genes.

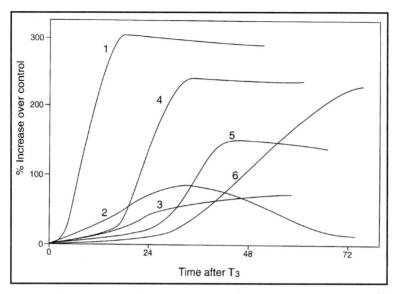

Fig. 4.2. Idealized curves summarizing the sequential stimulations of various biochemical activities of hepatocytes leading to increased mitochondrial respiration following a single injection of thyroid hormone (T_3) to young thyroidectomized rats. Inhibition of the first response by actinomycin D causes the loss of subsequent responses. Biochemical responses: 1, rapidly-labeled nuclear RNA in vivo; 2, run-on transcription by isolated nuclei (main product rRNA); 3, run-on transcription (main product non-ribosomal RNA); 4, accumulation of newly formed polyribosomes; 5, amino acid incorporation/mg RNA by polyribosomes in vitro; 6, amount of mitochondrial enzymes/mg protein and mitochondrial respiration. For details see Tata JR. Growth and developmental action of thyroid hormones at the cellular level. Handbook of Physiology 1974; Endocrinol III:469-78.

In all oviparous animals, the synthesis of the major egg proteins is under strict hormonal control. The ovary, via the neurosecretory and pituitary gonadotropic cascade system coordinates egg formation by activating the genes encoding egg white or coat and yolk proteins in the oviduct and liver, respectively[33-35] (see chapter 2, Fig. 2.6). Estrogen induces and maintains the synthesis of very large quantities of ovalbumin, conalbumin, ovomucoid and avidin in the oviduct of birds and that of yolk protein and lipid precursor vitellogenin in liver of birds and amphibia. In the laying chicken, as much as 1 g each of ovalbumin and vitellogenin are formed per day. In view of this highly hormone- and tissue-specific regulation of synthesis of massive amounts of protein, Korenman had initiated studies on the estrogen-ovalbumin system of the chick oviduct in order to explore the mechanism of action of steroid hormones. It was, however,

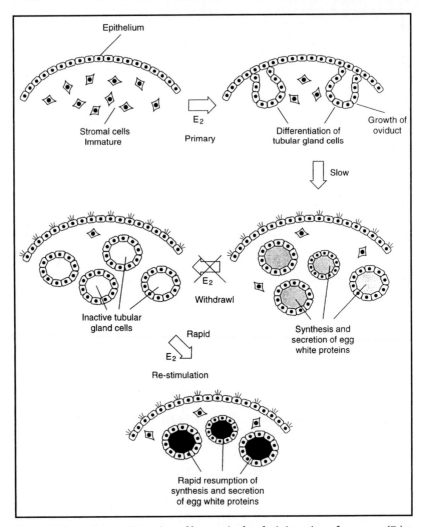

Fig. 4.3. Schematic representation of how a single administration of estrogen (E₂) to immature chick causes growth and differentiation of the oviduct to activate egg white protein genes in the newly formed tubular gland cells. The scheme also distinguishes between the slow, primary and rapid, secondary responses of the oviduct to hormonal treatment. Adapted from Schimke RT et al. Ovalbumin mRNA, complementary DNA and hormone regulation in chick oviduct. Karolinska Symposia on Research Methods in Reproductive Endocrinology 1973; 6:357-75.

O'Malley, followed later by the laboratories of Schimke and Chambon, who developed the system to be amenable for molecular investigation.[36-39]

    Figure 4.3 schematically illustrates how estrogen controls both the growth of the immature chick oviduct and the specific de novo induction of egg white protein synthesis in the single layer of secretory epithelial cells lining the lumen of the oviduct. Progesterone and glucocorticoids play a facilitative role, which was later explained on the basis of cross-regulation of receptor expression (see chapter 7). In the immature chicken, a single administration of estradiol-17β (or an agonist) causes a rapid hypertrophy and hyperplasia of the oviduct, from an organ of a few mg, to one of several g. The first response of the immature oviduct is cytodifferentiation to form tubular gland cells that are responsible for the synthesis of egg white proteins which is characterized by a lag period of up to 4 days, reaching a maximum at about 10 days. Withdrawal of the hormone causes the tubular gland cells to cease, making the egg white proteins by 21 days. If a second injection of E_2 is made at this time, egg protein synthesis resumes within a few hours and reaches a maximum level very rapidly. The rapidity of the response is due to the non-functioning tubular gland cells in which the ribosomes formed during primary stimulation are re-programmed by fresh mRNAs made during secondary stimulation. Besides activating the transcription of genes encoding the specialized or "luxury" proteins, estrogen, in common with all growth-promoting and developmental hormones, causes a massive and rapid accumulation of ribosomes, cellular membranes, secretory apparatus, etc.[34,37] The sequence of responses to estrogen (and indeed to all other growth and developmental hormones) is quite similar to that shown in Figure 4.2 for thyroid hormone.

Not surprisingly, the cytoplasm of the hormonally stimulated secretory cells of the oviduct are densely populated with polyribosomes programmed with mRNAs encoding egg white proteins soon after the administration of estrogen. It was therefore relatively easy to isolate transcripts encoding ovalbumin (and later, other egg white proteins) from the polysomes.[37,39,40] This led to the cloning of ovalbumin cDNA which in turn, made it possible to characterize the newly synthesized transcripts from the ovalbumin gene in intact oviductal cells, as well as, in run-on and run-off cell-free or reconstituted transcription assay systems. An important contribution of these developments was that it permitted a more precise understanding of the role of estrogen receptor (ER) and other nuclear hormone receptors.[32,41-43]

One of the major functions of estrogen in egg maturation in oviparous vertebrates is the coordination of the formation of egg proteins in the liver and oviduct. Thus, while the oviduct is engaged in synthesizing and secreting egg white protein, the polysomes in the liver are being reprogrammed to synthesize vitellogenin, such that 50-75% of polyribosomes are engaged in synthesizing this single protein at the height of its stimulation by estrogen.[33,44] A particular advantage of studying the vitellogenin gene in birds and amphibia is that this gene, which is normally silent throughout the lifetime of male animal, can be rapidly activated in the male liver by the hormone with similar characteristics as in the female. The large amphibian and avian vitellogenin gene is made up of 33 exons (~70 kb) and has been successfully used as a model for studying post-transcriptional processing of primary transcripts and maturation of particularly large mRNAs. The synthesis of vitellogenin is so intense that the cytoplasm of the avian and amphibian hepatocyte is swamped with the newly induced vitellogenin mRNA, causing the synthesis of albumin, which is the major protein synthesized in the liver (~7-8% of total) to be sharply reduced soon after hormonal induction.[44,45] The different rates of responses to primary and secondary hormonal stimulation of the immature chick oviduct (Fig. 4.3) is also reproduced in avian and amphibian liver for vitellogenin mRNA and that of amphibian egg coat proteins in the oviduct.[35,37,44–47] Fig. 4.4 illustrates the common features of the relatively slow, primary induction and the rapid, secondary induction of ovalbumin in chick oviduct and vitellogenin mRNA in organ cultures of adult, male *Xenopus* hepatocytes. This differential rate of activation of target genes during primary and secondary hormonal stimulation has often been termed as "memory" effect which has been explained as due to:

1) The higher level of estrogen receptor induced by estrogen during primary stimulation being responsible for the more rapid secondary response[48] (see also chapter 6).

2) Estrogen regulating differentially the stability of induced mRNAs.[49]

When genes encoding chicken egg white proteins other than ovalbumin were cloned, Palmiter[46] made the interesting observation that estrogen regulates the expression of these genes independently of one another. Even more impressive is the differential induction by estrogen of the individual members of the vitellogenin gene family in amphibian liver. Wahli's laboratory had established that in *Xenopus,* there are four closely related vitellogenin genes which were termed A1, A2, B1, B2 which are all expressed and regulated by estro-

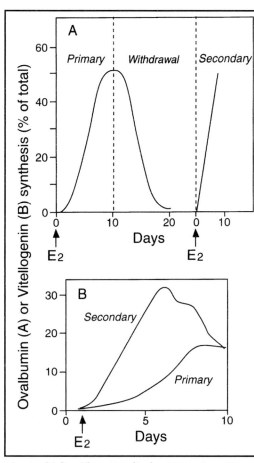

Fig. 4.4. "Memory" effect, as illustrated by the differential rates and magnitude of primary and secondary activation of A) ovalbumin gene in immature chick oviduct in vivo and B) vitellogenin genes in organ cultures of adult male *Xenopus* liver, following the administration of estradiol-17β (E_2). In A) 4-day old chicks were injected daily with 1 mg E_2 for 10 days of primary stimulation hormonal treatment was stopped (withdrawal) until ovalbumin synthesis had ceased, after which E_2 was again injected to produce the more rapid secondary response. In B) explants of naive (untreated) adult male Xenopus liver were cultured for 5 days in the presence of 1 μM E_2 to produce primary stimulation. For secondary stimulation, liver explants were similarly cultured but were obtained from adult male Xenopus treated 35 days earlier with a single injection of 1 mg E_2 when vitellogenin synthesis had ceased. Arrows indicate commencement of primary or secondary administration of estrogen. Data adapted from Schimke RT et al. Ovalbumin mRNA, complementary DNA and hormone regulation in chick oviduct. Karolinska Symposia on Research Methods in Reproductive Endocrinology 1973; 6: 357-75, and Green CD, Tata JR. Direct induction by estradiol of vitellogenin synthesis in organ cultures of male Xenopus laevis liver. Cell 1976; 7:131-9.

gen[50,51] (see Fig. 4.5). Using molecular probes that distinguish the individual transcripts Ng et al[52] showed in early *Xenopus* froglet liver that the four vitellogenin genes were induced and transcribed at different rates upon primary estrogen stimulation (see Fig. 4.6). These differences in the rate of transcription were more attenuated when vitellogenin genes were activated in adult male or female hepatocytes, both in vivo and in organ cultures and after secondary induction. Although the promoter regions of avian egg white proteins and amphibian vitellogenin genes exhibit many similarities within each group or family of genes,[43,50,51] there are sufficient differences that

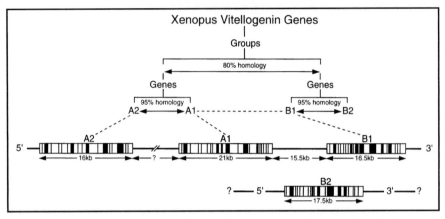

Fig. 4.5. Schematic representation of four vitellogenin genes of *Xenopus laevis*. These are classified as two closely related members of each of the A and B groups and termed A1, A2, B1 and B2. The figures as % denote the coding sequence homology between the two genes of each group and between the two groups (see refs. 50, 51).

could explain their non-coordinate regulation of transcription. Following, and in parallel with the development of the estrogen-ovalbumin gene system were other target genes activated by cognate hormones. Table 4.4 groups some well-known examples of hormone- and tissue-specific regulation of expression of target genes in different species.

What are the other benefits of research on the hormonal regulation of transcription of the ovalbumin and other genes? To cite only two impressive spin-offs: 1) The ovalbumin gene, following the discovery of the intron-exon structure of the mouse globin gene (see ref. 53), was among the first examples of split genes in eukaryotes,[40,54] and laid the foundations of our present-day understanding of post-transcriptional processing. 2) The possibility of setting up in vitro transcriptional systems with estrogen receptor and their target genes led to a more precise understanding of the action of anti-estrogens. Perhaps the best known of these are the non-steroidal antagonists hydroxytamoxifen and ICI164,384 which are extensively used in the therapy of hormone-dependent human breast cancer.[42]

Mechanism of Hormonal Control of Transcription

Progress in understanding how hormonal signals modulate the expression of their target genes is dependent on the state of our knowledge about eukaryotic gene transcription in general. This field has been rapidly moving over the last decade, particularly in the light of recent discoveries of transcription factors, so that it is im-

Table 4.4. Some examples of selective or specific hormonal induction of gene expression

Hormone	Species, Tissue	Gene Regulated de Novo	Induced	Rate of Transcription Regulated
Estrogen	Chick oviduct	Ovalbumin and other egg white proteins	+	+
	Chick liver	Vitellogenin	+	+
		VLDL	-	+
	Frog oviduct	FOSP-1	-	+
	Frog liver	Vitellogenin	+	+
Glucocorticoid	Rat liver	Tyrosine aminotransferase	-	+
Progesterone	Rabbit uterus	Uteroglobin	-	+
Testosterone	Rat, mouse prostate	PSA	-	+
Thyroid hormone	Rat liver, muscle	Mitochondrial dehydrogenases	-	+
	Rat heart	Myosin light chain	-	+
	Rat brain	Myelin basic protein	-	+
	Frog tadpole liver	Carbomyl phosphate synthetase	-	+
		Serum albumin	-	+
	Tadpole tail	Hydrolases, proteases	+/-	+
	Tadpole intestine	Stromelysin 3	-	+
	Tadpole skin	Adult keratin	+	+
Ecdysone	Insect larval cuticle	Tyrosine oxidases	-	+
	Larval salivary gland	Glue protein	+	+
Juvenile hormone	Larval fat body	Vitellogenin	+	+

VLDL, very low density lipoprotein; PSA, prostate-specific antigen; FOSP-1, frog oviduct-specific protein-1.

portant to first briefly consider what is currently known about the eukaryotic transcriptional machinery.

The Eukaryotic Transcriptional Machinery

Earlier studies on RNA polymerase II had suggested that the large number of subunits constituting this enzyme may have functional implications not only in the initiation, elongation and termination of transcripts, but also in the regulation of differential gene activity. However, with the cloning of promoters and the emergence of highly conserved sequences such as TATA and CCAT boxes and enhancers, and the emergence of common structural features of eukaryotic genes, it soon became obvious that the promoter played a central role in determining the modalities of transcription of a given protein-encoding gene.[27,53,55] As the isolation of gene promoters in-

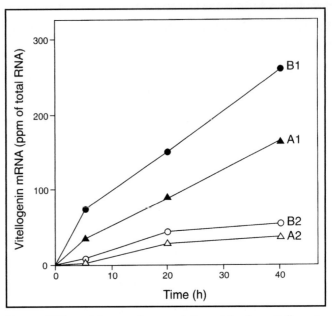

Fig. 4.6. Differential rates of accumulation of the four vitellogenin mRNAs in the livers of stage 66 *Xenopus* froglets as a function of time after treatment with 10^{-6} M estradiol-17β added to their water. For details see Ng WC et al. Unequal activation by estrogen of individual Xenopus vitellogenin genes during development. Dev Biol 1984; 102:238-47.

creasingly became an almost routine laboratory procedure, along with the development of transient cell transfection assays, the isolation of transcription factors also became progressively simple.[55-59] Transcription factors can be divided into two groups:

1) Those that are specific for a given gene or that function in a tissue- or development-specific manner.

2) Those that are general transcription factors irrespective of genes whose transcription they regulate or in the cells in which they function.

The transcription complex is made up of a large, continuously increasing number of proteins that interact among themselves and with the promoter DNA. Fig. 4.7 depicts some of the essential components of general transcription factors and other proteins that determine the initiation of transcription, which highlights both the highly specific protein-protein and protein-DNA interactions in a model originally proposed by Tjian,[58] and based on reconstitution of transcription complexes. These proteins are thought to interact simultaneously with DNA and among themselves, thus allowing for complexes to be formed between the distal and proximal regulatory elements. An important notion emerging from reconstitution experiments is that the temporal and positional order of protein-protein and protein-DNA interaction between transcription factors and the transcriptional machinery is highly critical for the generation of a functional transcription complex. Since nuclear hormone receptors are ligand-activated transcription factors, it is not surprising that many transcription factors interact with nuclear receptors and that they function synergistically.[60-62]

Many recent studies on transcription factors have focused on networking and integration of extracellular and intracellular signaling in order to explain how the activity of the transcription complex could be regulated.[63,64] As shown in the simplified model in Fig. 4.8, large protein molecules termed CBP and p300 would play a central role in forming bridges between nuclear hormone receptors and other transcription factors in order to activate hormonally regulated genes. These findings are based on the unexpected observation that nuclear receptors inhibit the activity of transcription factor AP-1 by competing for limited amounts of CBP (CREB binding protein)/p300 in cells. By interacting with multiple transcription factors and co-activators of nuclear receptors, some of which are indicated in Fig. 4.8, CBP/p300 serves to integrate multiple signaling pathways in the nucleus. It is important to realize that the functions of many transcription factors and nuclear receptors involved in such integration are themselves regulated by protein phosphorylation[65] (see chapter 3), thus setting up a network of membrane and nuclear receptor signal transduction pathways.

Nuclear Hormone Receptors as Ligand-Activated Transcription Factors

We have considered in chapter 3 how the specificity of a given nuclear receptor resides in the very precise DNA sequence of the cognate hormone response element (see also ref. 66). When deletion analysis of target gene promoters was undertaken, it soon became apparent that the relative positions of both the hormone response elements and the elements of other general transcription factors were major determinants in the activation of the target gene. A good example is the activation of the four *Xenopus* vitellogenin genes by estrogen. At the onset of their activation by estrogen in male *Xeno-*

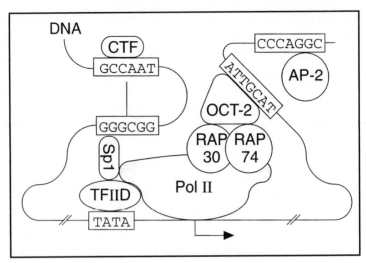

Fig. 4.7. A schematic view of how cis-elements of the promoter and enhancer regions of eukaryotic protein-coding genes transcribed by RNA polymerase II (Pol II) would form spatially defined complexes to facilitate transcription. The complexes would be generated by virtue of protein-protein interactions between transcription factors binding to DNA in a sequence-specific manner, as illustrated for TFIID, Sp1, CTF, AP-2 and OCT-2. RAP30 and RAP74 are components of the general transcription machinery. The interactions between transcription factors would effectively bring together the distal and proximal cis-elements to form a tight transcription complex. See ref. 56 for details.

pus hepatocytes, the four vitellogenin genes were found to be transcribed at unequal rates[52] (Fig. 4.6). When Wahli's laboratory determined the DNA sequence of the promoters of the four genes, they observed that the number and positions of the estrogen response elements (EREs) were not identical.[43,67] As shown in Fig. 4.9A, the most actively transcribed vitellogenin gene, i.e., B1, had three imperfect palindromic EREs whereas the least actively transcribed A2 gene had one perfect 13-nucleotide palindromic sequence GGTCACTGTGACC. The positions of the 1, 2 or 3 EREs, in the four genes were also not identical. This unequal distribution of the hormone response elements raises the possibility that the differential rates of transcription at the onset of activation of the silent genes may reside in the distribution of promoter sites that interact specifically with nuclear receptors. It is worth noting that the multiple chicken vitellogenin genes and another estrogen-activated gene in chicken liver (Apo-VLDL) also exhibit similar distribution patterns

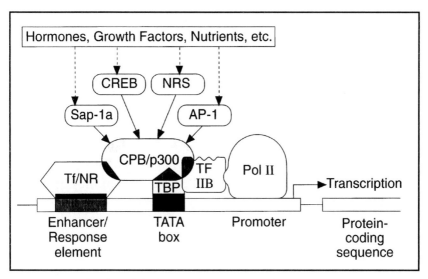

Fig. 4.8. A simplified model of a complex for RNA polymerase II (Pol II)-catalyzed transcription. A bridging protein such as CBP/p300 (CREB-binding protein) would closely contact sequence specific transcription factors (Tfs) and nuclear hormone receptors (NRs), TATA box-binding protein (TBP) and transcription factor IIB (TFIIB). The latter would not contact DNA but complex with Pol II. A factor such as CBP/p300 would form complexes with several other transcription factors without the involvement of DNA, such as NRs, CREB (cyclic AMP response element-binding protein), AP-1 and Sap-1a. The latter themselves are capable of binding DNA and their activities are modulated by phosphyration via MAPK and PKA (see Figs. 3.3, 3.4), thus allowing for the networking and integration of plasma membrane and nuclear signaling pathways (see refs. 63 and 64 for details).

of EREs in their promoters.[67]

Further sequence analysis of the *Xenopus* vitellogenin B1 gene revealed other sites in the promoter which are important for the ER to activate the transcription of the gene, shown as X, Y and NF-1 (see Fig. 4.9B). The best characterized of these is the response element for the NF-1-like transcription factor, which is known to be important for the regulation of the transcription of genes expressed in the liver.[43,68] Along with two other transcription factors (denoted as X and Y), NF-1 constitutes a basal transcription unit (BTU). It is suggested that in its chromatin configuration, the estrogen responsive unit (ERU), made up of EREs, would come in contact with the BTU to generate an active transcription complex by a 'looping' arrangement similar to that shown in Fig. 4.7. Perhaps the best example of an upstream regulatory factor interacting with nuclear receptor is that of chicken ovalbumin upstream promoter-transcription factor

(COUP-tf).[69] The COUP element, which was discovered in the course of studies on the regulation of ovalbumin gene by estrogen, is thought to be brought into close proximity of ER bound to the proximal ERE by virtue of its interaction with ER by a looping mechanism. Not included in Fig. 4.9 are other general transcription factors that would also act in concert with nuclear receptors, but which have been described in the literature.[62]

In most of the earlier studies, nuclear receptors were considered to play an activation or positive role in regulating transcription. However, it is becoming increasingly clear that they have both a positive and negative role to play. For example, based on the concept of interactions among transcription factors, Yamamoto's group[70] have described a 25 bp element in the proliferin gene promoter which conferred either a positive or negative glucocorticoid regulation of transcription. This sequence, termed a "composite" GRE bound both the glucocorticoid receptor and c-Jun and c-Fos which determined hormone responsiveness. c-Jun conferred a positive glucocorticoid effect on the composite GRE and a negative effect in the presence of high levels of c-Fos. Since c-Jun and c-Fos are components of the AP-1 transcription factor, this is yet another example of the integration of the extracellular and intracellular signaling pathways (see Fig. 4.8).

A simpler, dual positive and negative receptor-mediated transcriptional regulation is accomplished by the presence or absence of the ligand. Nuclear receptors have been commonly described as ligand-activated transcription factors, but a receptor acting as a repressor in the absence of the ligand had not been fully realized until thyroid hormone receptor was demonstrated to be a strong transcriptional repressor in the absence of thyroid hormone.[71,72] In one study a human TRβ mutant, associated with the clinical syndrome of generalized thyroid hormone resistance (GTHR), made it possible to locate the silencing function in the carboxyl terminal part of the receptor. A kindred S receptor was also shown to act as a constitutive repressor.[71] In the second study[72] unliganded TR was shown in vitro, using HeLa cell nuclear extracts and purified basal factor, to cause an active transcriptional repression which can only be relieved by the addition of T_3. The repression occurs at an early step during preinitiation complex assembly and it is independent of TBP-associated co-factors (see Fig. 4.8) involved in either basal repression or activator-dependent transcription.

More recently, co-activators and co-repressors of nuclear re-

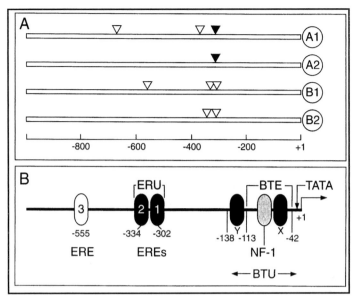

Fig. 4.9. Organization of the estrogen response elements (EREs) and possible other regulatory elements in the promoters of the Xenopus vitellogenin genes. A. Positions of 1, 2 or 3 perfect (▼) and imperfect (▽) consensus palindromic EREs (GGTCANNNTGACC) in the four Xenopus vitellogenin genes. B. Organization of the three EREs (1, 2 and 3) associated with sites for other trans-acting regulatory elements in the Xenopus vitellogenin B1 gene promoter. The basal transcription unit (BTU) near the TATA box is made up of 3 elements, X, Y and NF-1-like which are important for the functioning of the estrogen receptor interacting with the three EREs, the two closely positioned elements forming an estrogen responsive unit (ERU). It is suggested that in the natural configuration of the gene a looping of DNA would bring the ERU and BTU in close contact, necessary for the transcriptional activation of the vitellogenin gene by estrogen (see refs. 43, 67 for details).

ceptors, particularly retinoid and thyroid hormone receptors, have been purified with increasing emphasis being laid on the importance of synergistic and antagonistic actions resulting from their interactions.[73-76] Rosenfeld's group[73,74] have identified a large nuclear receptor co-repressor of 270 kDa (N-CoR) whose function suggests that the molecular mechanisms of repression by TR and RAR are analogous to those of the co-repressor-dependent transcriptional inhibition in yeast and *Drosophila*. They further show that DNA response elements can allosterically regulate RAR-co-repressor interactions, and thus determine whether the regulation of gene expression is to be positive or negative. Chen and Evans[75] have identified a co-repressor for retinoid and thyroid hormone receptors, termed silencing mediator (SMRT) of ~100 kDa whose association with RAR, RXR and TR is destabilized by the ligand (see Fig. 4.10). In the presence of T_3, SMRT would be dissociated from the TR-RXR heterodimer, allowing for the co-activator to enhance the activity of the basal transcriptional machinery. The transcriptional silencing by SMRTs is thus considered to be important in mediating hormone action. A cofactor for nuclear receptor has more recently been also described for homeobox genes which specify cell fate and positional identity in embryos.[77] A complex between fushi tarazu (Ftz) and a cofactor protein which is a member of the nuclear receptor family (Ftz-F1) has been isolated in *Drosophila*, the cofactor rendering the homeodomain protein to interact more strongly with its DNA binding site thereby enhancing the transactivation of genes under Ftz control. Undoubtedly more such co-activators and co-repressors of nuclear receptors will continue to be discovered in the future.

Chromatin Structure and Transcriptional Regulation by Nuclear Receptors

Complexities of transcriptional regulation inevitably lead one to consider the higher order of organization of transcription factors and nuclear receptors in the context of chromatin structure. Although there are many features of chromatin structure and transcription that remain to be explained, there is a growing realization that our understanding of hormonal regulation of gene expression would not be complete without taking into account how the nuclear receptors and accessory transcription factors are spatially organized into nucleosomal and chromosomal structures. The few examples below serve to illustrate this point, while the reader is referred to excellent recent reviews by Beato and his colleagues on how hormones influence chromatin structure and function.[61,78,79]

A major step forward was taken by Weintraub in the development of enzymatic techniques of studying the conformation of genes in transcriptionally active and inactive isolated nuclei and subnuclear preparations.[80] This advance made it possible later to demonstrate that in stimulating transcription some hormones also influenced the organization of the template or chromatin structure. Studies from Chambon's laboratory in the mid-1980s on the DNase I-hypersensitive regions of the chicken ovalbumin gene promoter defined how estrogen modifies the chromatin structure of this gene.[81,82] In one investigation these workers described the induction by estrogen of

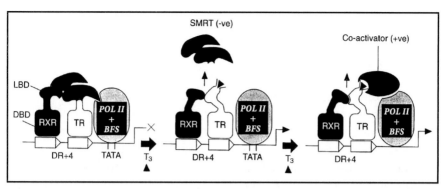

Fig. 4.10. Two-step model to explain how thyroid hormone (T$_3$) would activate transcription of its target gene by modifying the interaction between TR-RXR heterodimer, co-repressor (SMRT) and co-activator. In the first step SMRT would be associated with the unliganded TR-RXR heterodimer bound to the direct repeat+4 (DR+4) TRE and the basal transcriptional machinery (Pol II and BF), thus preventing any transcription. The binding of T$_3$ to TR would cause SMRT to be dissociated, leading to a low level of basal transcriptional activity. In the second step the ligand would facilitate the interaction between TR, co-activator and basal transcription machinery which would lead to a strong stimulation of transcription. DBD and LBD indicate DNA- and ligand-binding domains of the two receptors, respectively. BF is basal factors. See text and ref. 91 for details.

four hypersensitive sites, all or most of which disappear upon hormone withdrawal. The appearance and disappearance of these sites correlated well with the transcription of ovalbumin gene. In another study, they established that the four hormone-induced DNase I-hypersensitive sites were highly specific for oviduct chromatin. Interestingly, estrogen also altered chromatin structure at a polyadenylation and two putative transcription termination sequences at the 3' end of the gene. Micrococcal nuclease digestion of laying hen oviduct nuclei revealed that the nucleosomal pattern throughout the transcriptionally active ovalbumin gene was also altered by hormonal treatment.

In another hormonally induced egg protein gene system, estrogen stimulation also caused a highly gene- and tissue-specific change in the DNase I hypersensitivity of the *Xenopus* vitellogenin genes in hepatocytes.[83] Wahli's laboratory carried out a highly ingenious assembly of the *Xenopus* vitellogenin B1 gene into nucleosomes in a *Xenopus* oocyte extract,[84] such that transcription by RNA polymerase II of these DNA templates reconstituted into chromatin in a HeLa nuclear extract was 50-fold higher than with naked DNA template. Depletion of transcription factor NF-1 (see Fig. 4.9) in these extracts lowered transcription of the vitellogenin gene. Thus NF-1 participates, along with estrogen receptor, in the very efficient transcriptional potentiation of the promoter during nucleosomal assembly. In a collaborative study, Wahli's and Wolffe's groups determined the structural basis of this potentiation[85] in the light of the relative positions of EREs at ~530 with respect to the basal transcription unit at -138 to -42 in the *Xenopus* vitellogenin B1 promoter (Fig. 4.9). The model presented in Fig. 4.11 shows the DNA sequence-directed positioning of a nucleosome at -300 and -140 relative to the start of transcription. This positioning creates a static loop in which the two distal EREs at -302 and -334 are now brought close to the proximal basal sites upon activation of the gene by estrogen. These workers conclude that their positional nucleosome model is a good example of how a chromatin structure can have a positive effect on transcription (see also ref. 86).

Beato's laboratory have intensively studied how glucocorticoid hormone remodels the nucleosomal and chromatin organization and transcription of the mouse mammary tumor virus (MMTV)[79,87] Hormonal induction of MMTV requires the binding of GR to a complex of 4 GREs at -190 to -75 in a highly cooperative manner. This complex is precisely organized in phased nucleosomes and the process is accompanied by the binding of several transcription factors including NF-1 and OTF-1 (see also chapter 3, Fig. 3.12). In a later study they further show that MMTV promoter DNA is rotationally phased in intact cells. The nucleosome covering the MMTV promoter is neither removed nor shifted upon hormone induction and all associated transcription factors bind to the surface of the rearranged nucleosome. Most recently, several publications from many different laboratories have simultaneously demonstrated that an enzyme that catalyzes the removal of acetyl residues in histones is a key element in regulation of transcription (see ref. 88 for a review). Studying the transcription of different genes, these investigations come to almost the same conclusion that a chromatin-specific mechanism for transcriptional repression by a sequence-specific DNA-binding protein involves the recruitment of a deacetylase acting on core histones that is linked to DNA-bound repressors. What is particularly relevant here is the participation of SMRT and N-CoR (see preceding section and Fig. 4.10) in transcriptional repression. Evans and his colleagues[89] have demonstrated that histone deacetylase forms a repressor complex with SMRT which synergizes with retinoic acid to stimulate the transcription of hormone-responsive genes. Their findings are an

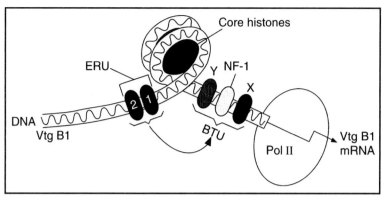

Fig. 4.11. Model to explain how the positional nucleosomal organization of the Xenopus vitellogenin B1 (Vtg B1) gene promoter would facilitate its transcription induced by estrogen. It is suggested that for transcription to proceed the estrogen response unit (ERU) composed of two EREs at -302 and -334 have to be brought in the proximity of the basal transcription unit (BTU) comprising the transcription factor NF-1 and two other elements X and Y between -42 and -138 (see Fig.4.9). The positioning of nucleosome at -300 to -140 from the transcription initiation site and the folding of the DNA around core histone, would create a static loop to facilitate the interaction between the ligand activated ERU and the proximal transcription factors (see ref. 85).

indication of the convergence of repression pathways, one of which involves nuclear receptors. They also raise the key question for future studies to resolve, namely how to fit these interactions into the higher order of organization of genes in chromatin, if not in the chromosomal context in vivo.

The few examples above reveal both the complexity of the structure-function relationship of transcription of genes at the higher order of organization. At the same time, they illustrate the different approaches available and progress made to date. Impressive as the conceptual and technical progress is in several areas, particularly as regards nuclear receptors, transcription factors, promoter elements, and chromatin structure, there are several questions relating to hormonal regulation of gene transcription that beg explanation. To cite only one: What is it that determines the high degree of tissue and gene specificity of hormones that regulate postembryonic development? Much still remains to be learned.

References

1. Tepperman J, Tepperman HM. Some effects of hormones on cells and cell constituents. Pharmacol Rev 1960; 12:301-53.
2. Yielding KL, Tomkins GM. Studies on the interaction of steroid hormones with glutamic dehydrogenase. Recent Prog Horm Res 1962; 18:467-85.
3. Tata JR. The search for the mechanism of hormone action. Pers Biol Med 1986; 29:184-204.
4. Riggs TR. Hormones and the transport of nutrients across cell membranes. In: Litwack G, Kritchevsky D, eds. Actions of Hormones on Molecular Processes. New York: Wiley, 1964.
5. Sutherland EW. Studies on the mechanism of hormone action. Science 1972; 177:401-8.
6. Cohen P. The role of protein phosphorylation in neural and hormonal control of cellular activity. Nature 1982; 296:613-20.
7. Berridge MV. Inositol lipids and calcium signaling. Proc R Soc Lond B 1988; 234:359-78.
8. Moncada S, Stamler J, Gross S et al. The Biology of Nitric Oxide: Part 5. London: Portland Press, 1996.
9. Knox WE, Auerbach VH, Lin ECC. Enzymatic and metabolic adaptations in animals. Physiol Rev 1956; 36:164-254.
10. Tomkins GM, Garren LD, Howell RD et al. The regulation of enzyme synthesis by steroid hormones: The role of translation. J Cell Comp Physiol 1965; 66:137-52.
11. Korner A. Hormones and protein synthesis in subcellular particles. Mem Soc Endocrinol 1961; 11:60-70.
12. Tata JR. The action of growth and developmental hormones. Biol Rev 1980; 55:285-331.
13. Tata JR. Regulation of protein synthesis by growth and developmental hormones. In: Biochemical Actions of Hormones. New York: Academic Press, 1970; 1:89-133.
14. Tata JR. Hormones and the synthesis and utilization of ribonucleic acids. Prog Nuc Acid Res Mol Biol 1966; 5:191-250.
15. Hamilton TH. Control by estrogen of genetic transcription and translation. Science 1968; 161:649-61.
16. Gorski J, Noteboom WD, Nicolette JA. Estrogen control of the synthesis of RNA and protein in the uterus. J Cell Comp Physiol 1965; 113:100-6.
17. Tata JR. Growth and developmental action of thyroid hormones at the cellular level. Handbook of Physiology 1974; Endocrinol III:469-78.
18. Clever U. Actinomycin and puromycin: effects on sequential gene activation by ecdysone. Science 1964; 146:794-5.
19. Williams-Ashman HG, Liao S, Hancock RL et al. Testicular hormones and the synthesis of ribonucleic acids and proteins in the prostate gland. Recent Prog Horm Res 1964; 20:247-301.
20. Clever U, Karlson P. Induktion von Puff-Veranderungen in den

Speicheldrusen-chromosomen von Chironomus tentans durch Ecdyson. Exp Cell Res 1960; 20:623-6.

21. Ashburner M. Patterns of puffing activity in the salivary gland chromosomes of Drosophila. VI. Induction by ecdysone in salivary glands of Drosophila melanogaster cultured in vitro. Chromosoma 1972; 38:255-81.

22. Pelling C. Puff RNA in polytene chromosomes. Cold Spring Harb Symp Quant Biol 1970; XXXV:521-31.

23. Russell S, Ashburner M. Ecdysone-regulated chromosome puffing in Drosophila melanogaster. In: Gilbert LI, Tata JR, Atkinson B, eds. Metamorphosis. Postembryonic Reprogramming of Gene Expression in Amphibian and Insect Cells. San Diego: Academic Press, 1996; 109-44.

24. Widnell CC, Tata JR. Studies on the stimulation by ammonium sulphate of the DNA-dependent RNA polymerase of isolated rat-liver nuclei. Biochim Biophys Acta 1966; 132:478-90.

25. Widnell CC, Tata JR. Stimulation of nuclear RNA polymerase during the latent period of action of thyroid hormones. Biochim Biophys Acta 1963; 72:506-8.

26. Roeder RG, Rutter WJ. Multiple forms of DNA-dependent RNA polymerase in eukaryotic organisms. Nature 1969; 224:234-7.

27. Chambon P. Summary: The molecular biology of the eukaryotic genome is coming of age. Cold Spring Harbor Symp Quant Biol 1978; XLII:1209-34.

28. Tata JR, Ernster L, Lindberg O et al. The action of thyroid hormones at the cellular level. Biochem J 1963; 86:408-28.

29. Oppenheimer JH, Samuels HH. Molecular Basis of Thyroid Hormone Action. New York: Academic Press, 1983.

30. Liao S. Cellular receptors and mechanisms of action of steroid hormones. Int Rev Cytol 1975; 41:87-172.

31. Ringold GM, Dobson DE, Grove JR et al. Glucocorticoid regulation of gene expression: mouse mammary tumor virus as a model system. Recent Progr Horm Res 1983; 39:387-424.

32. O'Malley BW, Vedeckis WV, Birnbaumer M-E et al. Steroid hormone action: the role of receptors in regulating gene expression. In: MacIntyre I, Szelke M, eds. Molecular Endocrinology. Amsterdam: Elsevier/North-Holland, 1977:135-50.

33. Tata JR, Smith DF. Vitellogenesis: a versatile model for hormonal regulation of gene expression. Recent Prog Horm Res 1979; 35:47-95.

34. O'Malley BW, Tsai M-J, Tsai SY et al. Regulation of gene expression in chick oviduct. Cold Spring Harbor Symp Quant Biol 1978; 42:605-15.

35. Tata JR, Lerivray H, Marsh J et al. Hormonal and developmental regulation of *Xenopus* egg protein genes. In: Roy AK, Clark JH, eds. Gene Regulation by Steroid Hormones IV. Springer-Verlag, 1989: 163-80.

36. O'Malley BW, McGuire WL, Kohler PO et al. Studies on the mecha-

nism of steroid hormone regulation of synthesis of specific proteins. Recent Prog Horm Res 1969; 25:105-53.

37. Schimke RT, Rhoads RE, Palacios R et al. Ovalbumin mRNA, complementary DNA and hormone regulation in chick oviduct. Karolinska Symposia on Research Methods in Reproductive Endocrinology 1973; 6:357-75.

38. Palmiter RD, Mulvihill ER, McKnight GS et al. Regulation of gene expression in chick oviduct by steroid hormones. Cold Spring Harbor Symp Quant Biol 1978; 42:639-47.

39. LeMeur M, Glanville N, Mandel JL et al. The ovalbumin gene family: Hormonal control of X and Y gene transcription and mRNA accumulation. Cell 1981; 23:561-71.

40. O'Malley BW, Roop DR, Lai EC et al. The ovalbumin gene: organization, structure, transcription and regulation. Recent Prog Horm Res 1979; 35:1-42.

41. Chambon P, Dierich A, Gaub M-P et al. Promoter elements of genes coding for proteins and modulation of transcription by estrogens and progesterone. Recent Prog Horm Res 1984; 40:1-39.

42. Parker MG. Nuclear Hormone Receptors. London: Academic Press, 1991.

43. Wahli W, Martinez E, Corthésy B et al. Cis- and Trans-acting elements of the estrogen-regulated vitellogenin gene B1 of *Xenopus laevis*. J Ster Biochem 1989; 34:17-32.

44. Deeley RG, Udell DS, Burns ATH et al. Kinetics of avian vitellogenin messenger RNA induction. J Biol Chem 1977; 252:7913-5.

45. Wolffe AP, Glover JF, Martin SC et al. Deinduction of transcription of *Xenopus* 74-kDa albumin genes and destabilization of mRNA by estrogen in vivo and in hepatocyte cultures. Eur J Biochem 1985; 146:489-96.

46. Palmiter RD. Regulation of protein synthesis in chick oviduct. I. Independent regulation of ovalbumin, conalbumin, ovomucoid and lysozyme induction. J Biol Chem 1972; 247:6450-61.

47. Wolffe AP, Tata JR. Coordinate and non-coordinate estrogen-induced expression of A and B groups of vitellogenin genes in male and female *Xenopus* hepatocytes in culture. Eur J Biochem 1983; 130:365-72.

48. Perlman AJ, Wolffe AP, Champion J et al. Regulation by estrogen receptor of vitellogenin gene transcription in *Xenopus* hepatocyte cultures. Mol Cell Endocrinol 1984; 38:151-61.

49. Shapiro DJ, Brock ML. Messenger RNA stabilization and gene transcription in the estrogen induction of vitellogenin mRNA. In: Litwack G, ed. Biochemical Actions of Hormones. Vol. XII, Orlando: Academic Press, 1985:139-72.

50. Wahli W, Dawid IB, Wyler T et al. Vitellogenin in *Xenopus laevis* is encoded in a small family of genes. Cell 1979; 16:535-49.

51. Wahli W, Ryffel GU. *Xenopus* vitellogenin genes. In: Maclean N ed.

Oxford Surveys on Eukaryotic Genes. Oxford: Oxford University Press, 1985; 2:97129.

52. Ng WC, Wolffe AP, Tata JR. Unequal activation by estrogen of individual *Xenopus* vitellogenin genes during development. Dev Biol 1984; 102:238-47.

53. Lewin B. Genes. IV. Oxford: Oxford University Press, 1996.

54. Royal A, Garapin A, Cami B et al. The ovalbumin gene region: Common features in the organisation of three genes expressed in chicken oviduct under hormonal control. Nature 1979; 279:125-32.

55. Roeder RG. The complexities of eukaryotic transcription initiation: regulation of preinitiation complex assembly. Trends Biochem Sci 1991; 16:402-8.

56. Mitchell PJ, Tjian R. Transcriptional regulation in mammalian cells by sequence-specific DNA binding proteins. Science 1989; 244:371-7.

57. Johnson PF, McKnight SL. Eukaryotic transcriptional regulatory proteins. Ann Rev Biochem 1989; 58:799-839.

58. Tjian R. The biochemistry of transcription and gene regulation. The Harvey Lectures 1996; 90:19-39.

59. Papavassiliou A. Transcription Factors in Eukaryotes. Berlin: Springer, 1997.

60. Schule R, Muller M, Kaltschmidt C et al. Many transcription factors interact synergistically with steroid receptors. Science 1988; 242:1418-20.

61. Beato M, Herrlich P, Schutz G. Steroid hormone receptors: Many actors in search of a plot. Cell 1995; 83:851-7.

62. Mangelsdorf DJ, Thummel C, Beato M et al. The nuclear receptor superfamily: the second decade. Cell 1995; 83:835-9.

63. Kamei Y, Xu L, Heinzel T et al. A CBP integrator complex mediates transcriptional activation and AP-1 inhibition by nuclear receptors. Cell 1996; 85:403-14.

64. Chakravarti D, LaMorte VJ, Nelson MC et al. Role of CBP/P300 in nuclear receptor signaling. Nature 1996; 383:99-103.

65. Jackson SP. Regulating transcription factor activity by phosphorylation. Trends Cell Biol 1992; 2:104-8.

66. Glass CK. Differential recognition of target genes by nuclear receptor monomers, dimers and heterodimers. Endocrin Rev 1994; 15:391-407.

67. Wahli W. Evolution and expression of vitellogenin genes. Trends Gen 1988; 4:227-32.

68. Chang T-C, Shapiro DJ. An NF1-related vitellogenin activator element mediates transcription from the estrogen-regulated *Xenopus laevis* vitellogenin promoter. J Biol Chem 1990; 265:8176-82.

69. Qiu Y, Tsai SY, Tsai M-J. COUP-TF. An orphan member of the steroid/thyroid hormone receptor superfamily. Trends Endocrinol 1994; 5:234-9.

70. Diamond MI, Miner JN, Yoshinaga SK et al. Transcription factor

interactions: selectors of positive or negative regulation from a single DNA element. Science 1990; 249:1266-72.

71. Baniahmad A, Tsai SY, O'Malley BW et al. Kindred S thyroid hormone receptor is an active and constitutive silencer and a repressor for thyroid hormone and retinoic acid responses. Proc Natl Acad Sci USA 1992; 89:10633-7.

72. Fondell JD, Roy AL, Roeder RG. Unliganded thyroid hormone receptor inhibits formation of a functional preinitiation complex: implications for active repression. Genes Dev 1993; 7:1400-10.

73. Hörlein AJ, Naar AM, Heinzel T et al. Ligand-independent repression by the thyroid hormone receptor mediated by a nuclear receptor co-repressor. Nature 1995; 377:397-404.

74. Kurokawa R, Soderstrom M, Hörlein A et al. Polarity-specific activities of retinoic acid receptors determined by a co-repressor. Nature 1995; 377:451-54.

75. Chen JD, Evans RM. A transcriptional co-repressor that interacts with nuclear hormone receptors. Nature 1995; 377:454-57.

76. Folkers GE, Burg VD, Saag VD. A role for cofactors in synergistic and cell-specific activation by retinoic acid receptors and retinoid X receptor. J Ster Biochem Mol Biol 1996; 56:119-29.

77. Yu Y, Li W, Kai S et al. The nuclear hormone receptor Ftz-F1 is a cofactor for the *Drosophila* homeodomain protein Ftz. Nature 1997; 385:552-555.

78. Beato M. Transcriptional control by nuclear receptors. FASEB J 1991; 5:2044-51.

79. Beato M, Candau R, Chavez S et al. Interaction of steroid hormone receptors with transcription factors involves chromatin remodelling. J Ster Biochem Mol Biol 1996; 56:47-59.

80. Weintraub H, Groudine M. Chromosomal subunits in active genes have an altered conformation. Science 1976; 193:848-56.

81. Kaye JS, Pratt-Kaye S, Bellard M et al. Steroid hormone dependence of four DNase I-hypersensitive regions located within the 7000-bp 5'-flanking segment of the ovalbumin gene. EMBO J 1986; 5:277-85.

82. Bellard M, Dretzen G, Bellard F et al. Hormonally induced alterations of chromatin structure in the polyadenylation and transcription termination regions of the chicken ovalbumin gene. EMBO J 1986; 5:567-74.

83. Gerber-Huber S, Felber BK, Weber R et al. Estrogen induces tissue specific changes in the chromatin conformation of the vitellogenin genes in *Xenopus*. Nucl Acids Res 1981; 9:2475-94.

84. Corthésy B, Leonnard P, Wahli W. Transcriptional potentiation of the vitellogenin B1 promoter by a combination of both nucleosome assembly and transcription factors: an in vitro dissection. Mol Cell Biol 1990; 10:3926-33.

85. Schild C, Claret F-X, Wahli W et al. A nucleosome-dependent static loop potentiates estrogen-regulated transcription from the *Xenopus* vitellogenin B1 promoter in vitro. EMBO J 1993; 12:423-33.

86. Wolffe AP. Regulation of Chromatin Structure and Function. Aus-

tin: RG Landes, 1994.

87. Truss M, Bartsch J, Schelberg A et al. Hormone induces binding of receptors and transcription factors to a rearranged nucleosome on the MMTV promoter in vivo. EMBO J 1995; 14:1737-51.

88. Pazin MJ, Kadonaga JT. What's up and down with histone deacetylation and transcription? Cell 1997; 89:325-8.

89. Nagy L, Kao H-Y, Chakravarti D et al. Nuclear receptor repression mediated by a complex containing SMRT, mSin3A, and histone deacetylase. Cell 1997; 89:373-80.

90. Green CD, Tata JR. Direct induction by estradiol of vitellogenin synthesis in organ cultures of male *Xenopus* laevis liver. Cell 1976; 7:131-9.

91. Mangelsdorf DJ, Evans RM. The RXR heterodimers and orphan receptors. Cell 1995; 83:841-50.

92. Liao S, Leininger KR, Sagher D et al. Rapid effect of testosterone on ribonucleic acid polymerase activity of rat ventral prostate. Endocrinology 1965; 77:763-5.

93. Gorski J. Early estrogen effects on the activity of uterine ribonucleic acid polymerase. J Biol Chem 1964; 239:889-92.

94. Kenney FT, Kull FJ. Hydrocortisone-stimulated synthesis of nuclear RNA in enzyme induction. Proc Natl Acad Sci 1963; 50:493-99.

95. Tata JR, Widnell CC. Ribonucleic acid synthesis during the early action of thyroid hormones. Biochem J 1966; 98:604-20.

CHAPTER 5

Metamorphosis: An Ideal Model for Hormonal Regulation of Postembryonic Development

Although the molecular analysis of the activation of ovalbumin and vitellogenin genes by estrogen in the immature chick oviduct and in the male avian and amphibian liver has significantly enhanced our understanding of hormonal control of specific gene expression, these egg protein genes are normally induced in fully differentiated adult tissues. There is therefore a need for a postembryonic developmental system in which the adult phenotype is hormonally induced by a process of gene switching in not fully differentiated tissues of free-living embryos. These requirements are admirably met by the postembryonic developmental process of vertebrate and invertebrate metamorphosis.

What is metamorphosis? The word, from its Greek derivation, simply means a change in form or shape. Biologists had long been intrigued by the dramatic and apparently spontaneous transformation of a caterpillar into a butterfly or that of a tadpole into a frog, both classical examples of metamorphosis. Early studies had indicated that these morphological changes were accompanied by profound alterations in chemical composition and biochemical functions. It was, however, not until the discovery that this postembryonic developmental process was dependent on endocrine secretions that it became possible to understand how metamorphosis was brought about and regulated. Before considering the molecular mechanisms underlying hormonal signaling controls that regulate this dramatic process of postembryonic development, it is worth taking into account the major characteristics of metamorphosis. The reader will find much useful detailed information in three volumes specifically dedicated to *Metamorphosis*, published at almost 15 year intervals.[1-3]

Hormonal Signaling and Postembryonic Development,
by Jamshed R. Tata. © 1998 Springer-Verlag and R.G. Landes Company.

Metamorphosis Is Under Obligatory Hormonal Control

In chapter 2, we have noted that unlike many developmental processes occurring during early embryogenesis, postembryonic development is not autonomous but is highly dependent on extracellular signals. Indeed, a salient feature of metamorphosis is that the process is obligatorily initiated and sustained by hormonal signaling. This endocrine control has been conserved through evolution in oviparous animals and is particularly dramatically illustrated by insect and amphibian metamorphosis. Figure 5.1 depicts the overall similarity of hormonal control in a moth and a frog, in which two groups of hormonal signals, released from specialized endocrine cells in response to environmental cues transmitted by neurosecretory cells in the brain, determine the onset, rate and completion of the postembryonic developmental process.

In insects (Fig. 5.1A), certain neurosecretory cells in the larval or pupal brain act on the *corpus allatum* to produce a hormone called prothoracicotropic hormone (PTTH) which acts on specialized cells of the prothoracic gland. The latter when stimulated by PTTH will in turn, secrete a group of steroid hormones termed ecdysteroids (the principal member being ecdysone) which induce metamorphosis.[4-9] At well-defined periods of the developmental progression leading to metamorphosis, different cells in the invertebrate larval and pupal brain stimulate the *corpus allatum* to produce and secrete a group of terpenoid compounds collectively termed juvenile hormone or JH.[4,6,7,10] As its name suggests, JH maintains the larva in a juvenile

Fig. 5.1. (Opposite) Simplified schemes depicting the hormonal regulation of insect and amphibian metamorphosis. a) In response to environmental cues (light, temperature, etc.), the neurosecretory cells (NC) in the brain of the moth larva stimulate the corpus allatum and prothoracic gland (with PTTH) to secrete juvenile hormone and ecdysone. The balance between these hormones determines the initiation and sustenance of a series of larval and pupal molts leading to the formation of the adult moth. b) Similarly, in the frog tadpole environmental factors trigger the sequential release of hypothalamic (CRF) and pituitary (TSH) hormones which stimulate the thyroid gland to synthesize and secrete thyroid hormones (T_4, T_3). Although not fully understood at the cellular level, exogenous prolactin prevents or slows down the action of T_4 and T_3. Other hormones (glucocorticoids, CRF) are also known to modulate the action of thyroid hormones. Abbreviations: PTTH, prothoracicotropic hormone; CRF, corticotropin releasing factor; TSH, thyroid stimulating hormone; T_4, T_3, L-thyroxine and 3,3',5-triiodo-L-thyronine; GC, glucocorticoid hormone.

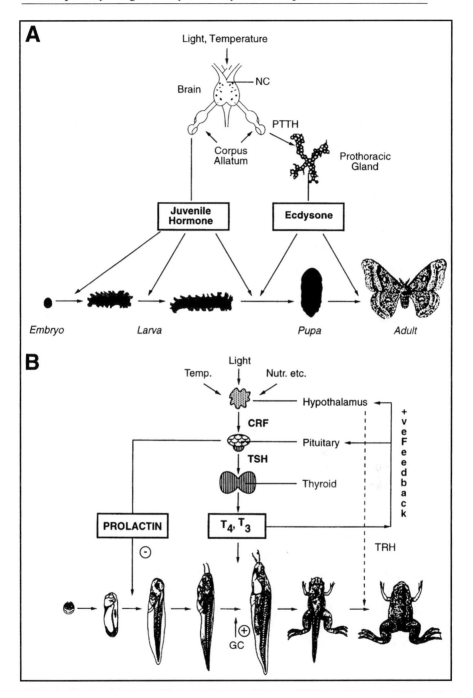

state and prevents metamorphosis. This juvenilizing action is essential in determining the timing of initiation of the program for further development of growing but not fully differentiated tissues. Although the intracellular sites and mechanisms of action of juvenile hormone are not fully understood, it is significant that JH will counteract many biochemical actions of ecdysteroids at the cellular level.[11] In contrast, much is now known about nuclear receptor and transcriptional activation by ecdysteroids (see chapters 3 and 4). Perhaps more interesting is the fact that titres of JH go up and down sharply, almost reciprocally with those of ecdysone, illustrated in Figure 5.2A for the tobacco horn worm *Manduca*.[7] This pulsatile hormonal signaling has the effect of producing multiple larval and pupal molts in many insects, each repeated with a similar counteracting interaction between JHs and ecdysteroids.

In amphibia (and other metamorphosing vertebrates), it has been known since the discovery by Gudernatsch in 1912[12] of the precocious induction of metamorphosis in frog tadpoles that the process is under hormonal control.[1,13,14] With the recognition of the central role played by the hypothalamus-pituitary-thyroid axis in vertebrates, the link between environmental signals and the initiation of metamorphosis could also be traced to the central nervous system through the intermediary of hypothalamic hormones TRH (thyrotropin releasing hormone), CRF (corticotropin releasing factor) and TSH (thyrotropic hormone) made in the pituitary (see Fig. 5.1B). The concept of a positive feedback loop between thyroid hormones (T_4, T_3) and TSH at the onset of amphibian metamorphosis, first proposed by Etkin, is in marked contrast to the general notion of negative feedback loops operating between the hypothalamus-pituitary axis and peripheral endocrine glands.[1,15] Although thyroid hormone (TH) is the only obligatory signal for the initiation and completion of amphibian metamorphosis, other hormones and factors can modulate the onset and progression of metamorphosis. These include glucocorticoid hormone, as well as CRF and ACTH (adrenocorticotropin), which can accelerate TH-induced metamorphosis both in intact tadpoles and isolated tissues.[16] Amphibian metamorphosis is known to be sensitive to photoperiodicity and circadian rhythms. It is slowed down in the dark and by exogenous melatonin which is principally produced in the pineal gland.[17] Many investigators have reported that exogenous mammalian and amphibian prolactins (PRL) will prevent both natural and TH-induced metamorphosis in many different species of tadpoles.[16,18-21] Intriguingly, circulating PRL levels tend to increase in *Xenopus* and bullfrog tadpoles during and after metamorphosis, suggesting possible other roles for this hormone in adults not yet fully defined.[16] Nevertheless, PRL has been shown to inhibit equally T_3-induced growth and differentiation of the *Xenopus* tadpole limb bud and the regression of the tail directly in organ culture.[22] Although the suggestion that PRL can be considered as a vertebrate juvenile hormone is debatable, as will become apparent later, this hormone can be a useful tool in exploring the mechanism of action of TH in regulating amphibian metamorphosis. In contrast to the interrupted pattern of metamorphosis in insects (Fig. 5.2A), in vertebrates it is an uninterrupted larval-adult transition (Fig. 5.2B) and follows the circulating level of thyroid hormone.[23]

Simple exposure of insect larvae to ecdysone or of amphibian tadpoles to TH can precociously induce normal metamorphosis. Conversely, withholding these hormones by surgical or chemical ablation of the respective endocrine gland will prevent further development or metamorphosis until such time as the hormone is replaced.[1,13] Precocious induction of metamorphosis by simple hormonal manipulation as an experimental approach has greatly contributed to our understanding of the physiology and biochemistry of postembryonic development.

Hormonal Activation Is Direct and Local

An important feature of the action of metamorphic hormones is that it is not systemic but direct and local. Early studies on ligation of insect larvae and local application of hormones to different parts of the frog tadpole established this point.[1,7,13,24] The examples shown in Fig. 5.3 illustrate this principle. A perfect illustration is when L-thyroxine (T_4) dissolved in cholesterol (to prevent its rapid diffusion) was applied externally to one side of the tadpole tail fin and the base alone to the other, only the site at which T_4 was applied underwent local regression (Fig. 5.3a). Similar results were obtained when only the tadpole eye to which T_4 was applied locally, contained rhodopsin (adult) and not the control eye which had porphyropsin (larval) as the visual pigment (Fig. 5.3b). The porphyropsin ⇨ rhodopsin transition, along with that of larval-adult hemoglobin, are among the major biochemical processes of gene switching during amphibian metamorphosis.[25] Ligation experiments in insects, as illustrated

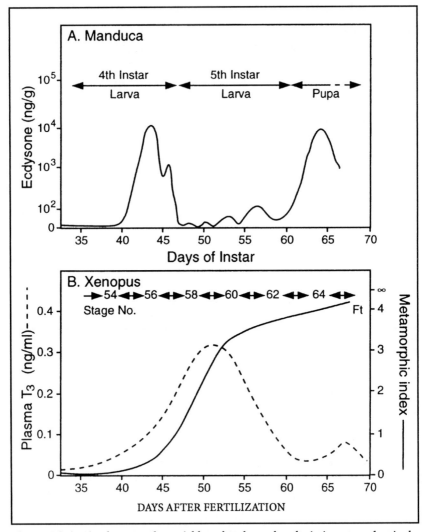

Fig. 5.2. Distinction between the serial larval and pupal molts in insects and a single transition from larval to adult forms in amphibian metamorphosis. The idealized curves depict a) the pulses of ecdysone released (ng/g of whole body) just preceding each of the multiple larval and pupal molts in *Manduca sexta* and b) the single burst of T_3 (ng/ml plasma) just before a continuous transition from the larval to the adult state in *Xenopus laevis*. Metamorphic Index is ratio of hind limb : tail length. Data for *Manduca* adapted from ref. 7.

for the horn worm *Manduca* in Fig. 5.3c, whereby ecdysteroids produced in the prothoracic gland caused the top half of the larva to turn into an adult while the bottom half retains its larval form and composition. The elegant experiments of Becker in 1962 on ligation of *Manduca* larvae extended these observations to gene puffing, a

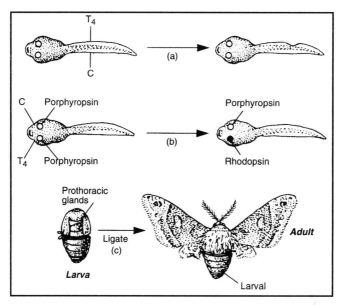

Fig. 5.3. Some examples illustrating the principle of direct and local, as opposed to systemic, effects of hormones in inducing metamorphosis. a) Thyroxine (T_4) dissolved in cholesterol was applied directly to one side of a tadpole tail fin (C = control); after a few days only a small region of the side to which the hormone was applied underwent resorption. b) Tadpole eyes contain the larval visual pigment *porphyropsin*, whereas the adult frog has the terrestrial visual pigment *rhodopsin*. The eye to which thyroxine was applied contained rhodopsin after a few days, whereas the control eye receiving only cholesterol base still contained porphyropsin. c) Just before metamorphosis, an insect larva is ligated below the prothoracic glands, which synthesize the hormone ecdysone. After a few days when endogenous ecdysone is secreted the top half of the larva turns into an adult while the bottom half still retains its larval form and composition.[24]

salient feature of insect metamorphosis in tissues exhibiting polytenic chromosomes, which similarly pointed to the local effects of ecdysone (see ref. 5).

Equally convincing proof of local effects is provided by organ culture of larval tissues exposed to the metamorphic hormones. In the mid-1960s it was possible to show that T_3 added to organ cultures of early pre-metamorphic *Xenopus* tadpole tails caused them to undergo regression with the same morphological and biochemical criteria seen during natural metamorphosis[26] (see chapter 8). Later, as illustrated in Fig. 5.4, organ cultures demonstrated the local

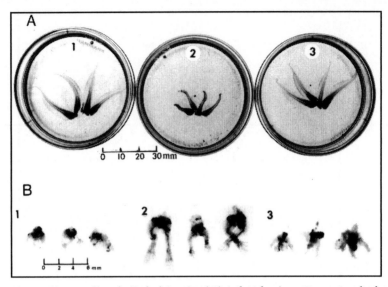

Fig. 5.4. Direct action of triiodothyronine (T_3) and prolactin on Xenopus tadpole a) tails and b) hind limb buds maintained in organ culture. Tails and limb buds were taken from the same stage 54/55 tadpoles and cultured for 8 days with and without 2.2 nM T_3 and 0.2 units/ml of PRL added one day after T_3. 1: control; 2: T_3; 3: T_3 and PRL. These photographs are also meant to show the strong inhibition by exogenous PRL of T_3-induced regression as well as morphogenesis during amphibian metamorphosis. Other details can be found in ref. 22.

and direct effects of T_3 for both the cell death response of the *Xenopus* tadpole tails and the induction of morphogenesis of cultured hind limb buds removed from the same larvae.[22] These studies further showed that the direct tissue effects were not restricted to the inductive actions of T_3, but were also applicable to the anti-metamorphic effects of prolactin. In 1971 Ashburner demonstrated that the in vivo temporal pattern of chromosomal puff appearance and regression in *Drosophila* larvae could be perfectly mimicked in organ cultures of larval salivary glands.[27,28] Upon the addition of 20-hydroxyecdysone to cultures of salivary glands of third instar larvae, he found the intermolt puffs to regress, and a small set of early puffs to be rapidly induced at rates characteristic for each puff in vivo. Interestingly, if after some hours the glands were washed free of the hormone and then re-exposed to ecdysone, the hormone-free period mimicked the in vivo situation in which a drop in ecdysone titre occurs between the prepupal and pupal stages.

Another important feature of local effects of hormones worth considering is the fact that the tissue responses are independent of the location of the target cells. Early transplantation experiments

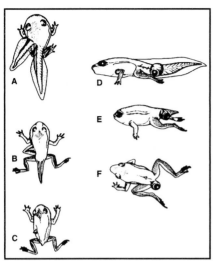

Fig. 5.5. Early transplantation experiments showing that metamorphic responses to thyroid hormone were inherent in the tissues of the amphibian tadpole independent of those of neighboring tissues. Tails of *Rana pipiens* tadpoles were transplanted on the lower side of the head (a, b, c) while one eye was transplanted at the proximal end of the tail (d, e, f). When the tadpoles were exposed to L-thyroxine (T_4) added to their water (b, c, e, f) the transplanted tissue underwent the same morphological and biochemical transformations as did the same tissues in their normal locations. For further details see ref. 13.

made this point quite clear.[13] As shown in Fig. 5.5, in frog tadpoles in which the tail was transplanted just below the head and then allowed to develop naturally, or exposed to exogenous thyroid hormone, the transplanted tail proceeded to regress just as the normal tail although the tissue surrounding the transplant did not. Conversely, the eye transplanted on the tail underwent biochemical differentiation and morphogenesis to the adult phenotype, whereas the surrounding tail tissue regressed and ultimately disappeared. These demonstrations of direct hormonal responses and tissue positional independence point to the important fact that the hormone merely serves to initiate a dormant developmental program laid down at an early stage in postembryonic development.

Metamorphic Hormones Initiate Diverse Tissue-specific Genetic Programs

Since the 1930s much information has accumulated about the diversity and extensiveness of morphological and biochemical changes in the insect and amphibian larvae in response to ecdysteroids and thyroid hormones, respectively.[2,3,7,10,29,30] Tables 5.1 and 5.2 list some of the well-known responses of various insect and amphibian larval tissues to their respective metamorphic hormones. Most noticeable in these two tables is that, although different genes constitute the responses in different cell types, there is a common feature underlying the hormonally regulated postembryonic developmental process, namely the acquisition of the adult phenotype in anticipation of a change in the environment of the organism. Direct

Table 5.1. *Some morphological and biochemical changes characteristic of metamorphosis induced by ecdysteroids in insect larvae*

Tissue	Morphological change	Biochemical response
Brain	Major restructuring and death of neurons and changes in sensory systems	Many gene products for sexual differentiation
Salivary gland	Massive regression; chromosomal puffs	Production of hydrolases, glue protein, alcohol dehydrogenase
Fat body	Resorption and re-organization	Induction of fat body-specific proteins
Epidermal cells	Cuticle formation and pigmentation	Induction of DOPA decarboxylase, polyphenol oxidase, cocoonase
Imaginal discs	Remodeling	Activation of numerous genes
Wing buds	De novo morphogenesis, scale and pigmentation	Cell proliferation, enzymes for tyrosine oxidation

evidence that gene switching during metamorphosis is anticipatory rather than an adaptational response is provided by the precocious hormonal induction of metamorphosis. An injection of TH or ecdysone to early tadpoles or insect larvae, respectively, leads to the activation of the developmental program. In all instances, the biochemical changes are preceded by an early burst of RNA synthesis, which is essential for the later changes in phenotype as shown by experiments with transcriptional inhibitors. This is seen in Figs. 5.6 and 5.7 for the induction by ecdysone of chromosomal puffs, some of which contain newly synthesized mRNA encoding salivary protein in the *Chironomus* larvae and by T_3 of urea cycle enzymes (carbomyl phosphate synthetase, ornithine transcarbamylase, arginosuccinate synthetase) and serum albumin in early bullfrog (*Rana catesbeiana*) tadpole liver, respectively (see also refs. 5,6,31-33).

The second striking feature of hormonally regulated postembryonic development is that no two tissues or groups of cells exhibit the same hormonal response, which range as widely as de novo morphogenesis, functional reprogramming and total tissue regression. To consider a few examples, the invertebrate and vertebrate larval CNS is a major hormonal target with a wide variety of changes taking place during metamorphosis in both its anatomical and functional characteristics. Similarly, the epidermal tissue or skin, and the liver or fat body undergo genetic reprogramming leading to the ac-

Table 5.2. *Diversity of morphological and biochemical responses during thyroid hormone-induced amphibian (anuran) metamorphosis*

Tissue	Response Morphological	Biochemical
Brain	Re-structuring, axon guidance, axon growth, cell proliferation and death	Cell division, apoptosis and new protein synthesis
Liver	Re-structuring, functional differentiation	Induction of urea cycle enzymes and albumin; Larval to adult hemoglobin gene switching
Eye	Re-positioning; new retinal neurones and connections; lens structure	Visual pigment transformation (porphyropsin → rhodopsin); β crystallin induction
Skin	Re-structuring; skin granular gland formation; keratinization and hardening; apoptosis	Induction of collagen, 63 kDa (adult) keratin and magainin genes; induction of collagenase
Limb bud, lung	De novo formation of bone, skin, muscle, nerves, etc.	Cell proliferation and differentiation; chondrogenesis
Tail, gills	Complete regression	Programmed cell death; induction and activation of lytic enzymes (collagenase, nucleases, phosphatases, matrix metalloproteinases); lysosome proliferation
Pancreas, intestine	Major tissue re-structuring	Reprogramming of phenotype, acquisition of new digestive functions, induction of trypsin and other proteases, fatty acid binding protein, sonic hedgehog (morphogen), stromelysin-3.
Immune system	Re-distribution of cell populations	Altered immune system and appearance of new immunocompetent components
Muscle	Growth and differentiation; apoptosis	Induction of myosin heavy chain

See refs. 20,21,33,37 for further details.

quisition of new morphological and biochemical characteristics, such as the appearance of digestive enzymes in the insect fat body and that of urea cycle enzymes and serum albumin in the amphibian liver, the epidermal sclerotization and enzymes responsible for the process in the formation of the adult insect cuticle, the keratinization of the amphibian larval skin as a result of the switching on of adult keratin gene, to cite only a few. Among the most dramatic

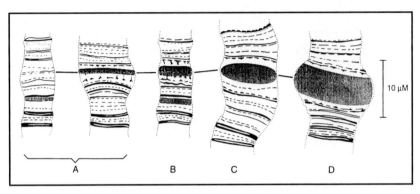

Fig. 5.6. Puffing of a specific region 1-17-β of one of the polytenic chromosomes of the salivary gland of *Chironumus* larva induced precociously by the injection of ecdysone. During natural or artificially hormone-induced metamorphosis, the salivary gland, which consists of only about 50 cells, undergoes regression after a series of sequential chromosomal puffs. With the availability of molecular techniques it was possible to show that the puff (here represented by the grey shading) was largely composed of newly synthesized RNA among which specific newly activated gene products could be identified. a, b, controls; c, d, ecdysone-injected larvae. Based on the work of U Clever. Dev Biol 1963; 6:73-98. (see also ref. 5).

morphological and biochemical changes are the simultaneous emergence of wings and limbs and total or almost total loss of larval tails, gills and the digestive system in insects and amphibia. Indeed there are few, if any, cells in the larval or pupal cells that escape the impact of metamorphic hormones. Thus, tissue-specific gene switching is central to hormonal signaling and postembryonic development. It is also worth re-emphasizing that the hormone does not determine the developmental program, but serves to initiate it.

An example of gene switching by turnover and replacement of cellular populations during metamorphosis is provided by the switch from larval to adult hemoglobin in erythroid cells of *Xenopus* larvae. At one time it was thought that the switch to express adult hemoglobin genes occurred in the same nucleated erythrocyte that carried larval hemoglobin prior to the onset of metamorphosis. However, immunofluorescence with specific antibodies led Weber and colleagues[34,35] to demonstrate that the larval hemoglobin carrying cells were progressively replaced by those expressing adult Hb following the onset of metamorphosis (Fig. 5.8).

Following the advent of molecular techniques to identify transcripts of individual genes, a large number of genes activated by TH and ecdysteroids in various larval amphibian and insect tissues during natural and induced metamorphosis have been identified.[5,19,33,35-38]

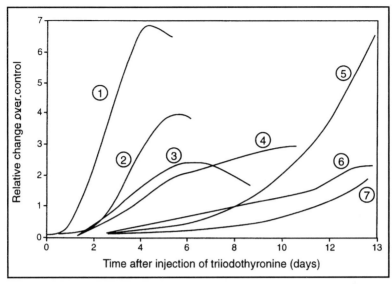

Fig. 5.7. Sequence of biochemical events in the hepatocytes of bullfrog (*Rana catesbeiana*) tadpoles following the injection of T_3 to induce metamorphosis precociously. 1, total nuclear RNA synthesis; 2, accumulation of newly synthesized RNA and polyribosomes (including mRNA) in the cytoplasm; 3, change in the rate and pattern of formation of microsomal membranes; 4, overall rate of protein synthesis per unit ribosomal RNA; 5, newly synthesized urea cycle enzymes; 6, increased rate of formation of mitochondria; 7, de novo synthesis and secretion of serum albumin. Note that the sequence resembles that in mammalian liver stimulated by thyroid hormone (Fig. 4.2). See ref. 24 for further details.

These include such genes as serum albumin, adult hemoglobin and carbomyl phosphate synthetase in *Xenopus* tadpole liver and alcohol dehydrogenase and glue protein in *Drosophila* larvae. In the laboratories of Brown and Shi, the analysis of gene switching has been extended to the characterization of genes that are silenced or downregulated during amphibian metamorphosis.[37,38] Table 5.3 lists some of the up- and down-regulated genes in the limb bud (de novo morphogenesis), intestine (partial regression) and tail (total cell death) by TH administered to premetamorphic *Xenopus* tadpoles. Many unidentified up- and down-regulated genes are not included in this table. Of particular interest are genes that can be classified as 'early' or 'direct response' genes (activated in the absence of protein synthesis) since it is likely that their products may play a causal role in the cascade of regulatory elements leading to tissue-specific biochemical and morphological changes listed in Tables 5.1 and 5.2. Also noteworthy is the fact that many transcription factor genes have been

Table 5.3. *Some up- and downregulated genes in* Xenopus *tadpole tissues in response to thyroid hormone* (T_3)

Tissue	No. of genes Upregulated	No. of genes Downregulated	Upregulated direct response genes
Hind limb	14	5	Heat shock protein
			Zn finger (E4.BP4)
Intestine	22	1	NF-1
			Na^+/PO_4^{3-} Cotransporter
			bZip (E4BP4)
			Stromelysin-3
Tail	35	10	Zinc finger (BTEB)
			Stromelysin-3
			Iodothyronine deiodinase

See refs. 37,38 for further details.

identified. The most significant in this class of genes are those encoding receptors for TH (and for ecdysteroids in insects). Its significance will be discussed in detail in the next chapter.

From early studies on the activation of the insect larval prothoracic gland or the amphibian tadpole's thyroid gland by the cascade of neurosecretory hormones, it had already emerged that metamorphosis would occur soon after these glands begin to secrete ecdysteroids or thyroid hormones (see Fig. 5.2 and refs. 1,2). At the same time, ligation, ablation or chemical inactivation of prothoracic and thyroid glands in early development, followed some time later by administration of exogenous ecdysteroid or TH clearly established that all the larval tissues were competent to undergo metamorphic changes before these glands became active. These observations raised the important question as to how early in development would the larval tissues be capable of responding to the metamorphic hormones, a question highly relevant to the developmental expression of their receptors. A systematic study in which different stages of *Xenopus* embryos and tadpoles were exposed to TH gave a clearcut answer to the above question.[39] Measurement of a number of morphological and biochemical responses revealed that the competence to respond to T_3 was established as early as stage 44 or 45, i.e., a week after fertilization of the egg (see Fig. 5.9). By stage 47 or 48, which is

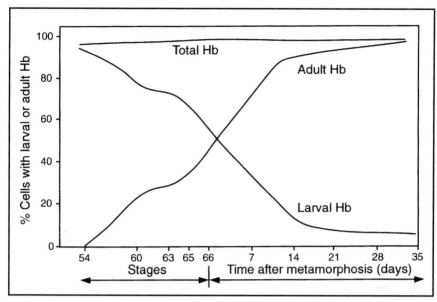

Fig. 5.8. An example of gene switching by cell replacement during amphibian metamorphosis. Larval and adult hemoglobin (Hb) cells were monitored immunochemically in the blood of Xenopus larvae at the onset, during and upon completion of metamorphosis. Adapted from refs. 34, 35.

about 2 weeks later, all the responses are clearly more evident. Natural metamorphosis begins at stage 53 or 54, which depending on external conditions, can be about 2 months after fertilization. As shown in Fig. 5.2B, TH first appears in the tadpole blood at about this time. The rapid build-up and decline in circulating TH temporally matches the onset and completion of natural metamorphosis. Thus, the acquisition of the response can be seen several weeks before the first detection of the hormone. Similar observations of early competence of response have been recorded for insect metamorphosis.

The above issue of early developmental competence is better illustrated by using molecular markers that specify the adult or post-metamorphic phenotype. Serum albumin is a major hepatic gene that is switched on during thyroid hormone-induced metamorphosis in all anurans (Table 5.2). If early pre-metamorphic *Xenopus* tadpoles (at stages up to 45) are exposed to T_3 for 3 days, the albumin gene is not expressed in response to the hormone (Fig. 5.10). However, at later stages, as for example at stage 48 shown in Fig. 5.10, the albumin gene is precociously induced by the hormone.[40] These findings lead to the conclusion that the functional receptor must be in place considerably before the hormonal signals impinge on the tissues during normal development and highlight the importance of considering the expression of receptor genes before and during metamorphosis (see chapter 6).

Programmed Cell Death Is an Important Feature of Metamorphosis

Upon the onset of natural or hormone-induced metamorphosis, the insect and amphibian larva exhibits a rapid and substantial loss of cells in many tissues which continues until the process is completed. This loss of cell number and total mass of the organism is manifested as the loss of entire organs, as is the case for tadpole tail and gills. However, this postembryonic developmental cell death is not restricted to total tissue regression, but can be quite extensive in tissues that are morphologically and functionally re-structured, such as the brain, gut and salivary glands in insects and the intestine, pancreas and skin in amphibia. These tissues will undergo further development and can exhibit substantial cell proliferation as metamorphosis reaches completion.

Earlier studies to explain hormonal induction of tissue regression during metamorphosis were based on such processes as macrophage infiltration, lysosomal expansion or activation of lytic enzymes.[41-46] However, TH or ecdysteroids fail to directly activate these enzymes (proteases, nucleases, phosphatases) in their latent forms, as did other agents that activate lysosomal enzymes. These lysosomal activators fail to induce tissue regression characteristic of metamorphosis. This has raised the possibility that the hormonal elevation of lytic enzyme activity in larval tissue programmed for regression was caused by a selective enhancement of the synthesis of some or all of them. Studies on induction of regression in organ cultures of tadpole tails showed that T_3 simultaneously augmented the amount of several lytic enzymes and the protein and RNA synthesizing activity of the tadpole tails. The use of inhibitors of RNA and protein synthesis demonstrated the requirement of newly synthesized RNA and protein to initiate cell death during metamorphosis and other postembryonic developmental processes.[43,47,48] Programmed cell death during metamorphosis is discussed in detail in chapter 8.

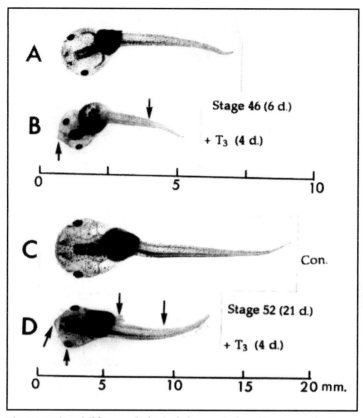

Fig. 5.9. Major visible morphological changes induced in very early stages of Xenopus tadpoles by thyroid hormone. (A,C) Control animals on days 10 and 25 after fertilization, respectively. (B,D) Tadpoles at 6 and 21 days after fertilization, respectively, treated with 4×10^{-9} M T$_3$ and photographed 4 days later. Arrows indicate the regression of the tail, the appearance of a beak-like structure, the altered positioning of eyes, and swelling of hind-limb buds. Note that tadpoles at later stages were more sensitive to the hormone. These morphological changes are accompanied by several biochemical responses (see ref. 39 for further details).

Neotenic Amphibia and Diversity of Metamorphosis

The above examples of insect and amphibian metamorphosis are among the most dramatic morphogenetic transformations during postembryonic development. Less striking or externally visible metamorphic changes are, however, quite widespread in vertebrates and invertebrates. Thus, larval and adult forms can be distinguished in coelenterata, flat worms, molluscs, crustacea, echinoderms, cyclostomes, eels, teleost fishes, etc. Metamorphic transitions can

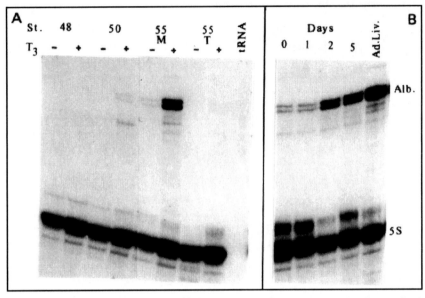

Fig. 5.10. Early developmental competence of Xenopus tadpoles to respond to T$_3$, as judged by the activation of albumin gene. A. Batches of tadpoles at stages from 48 to 55 were treated (+) or not (-) for 4 days with 2 nM T$_3$. RNA was extracted and albumin mRNA determined by RNase protection assay, using Xenopus 5S RNA as an internal control. Low level signals for albumin mRNA were visible in tadpoles from stage 50 onwards. B. Time-course of accumulation of albumin mRNA in tadpoles of stage 54 exposed to T$_3$ for 0-5 days. For comparison, an equivalent amount of adult liver RNA (Ad. Liv.) was assayed at the same time. The position of albumin mRNA (Alb) and 5S RNA (5S) are indicated. For other details see Baker BS, Tata JR. Accumulation of proto-oncogene *c-erb*-A related transcripts during *Xenopus* development: association with early acquisition of response to thyroid hormone and estrogen. EMBO J 1990; 9:879-85.

range from only partial or attenuated transitional forms to extremely brief or delayed periods of metamorphosis.[25,49-54] For example, some invertebrate larval forms last for a few minutes in contrast to the sea lamprey which lives for several years as a larva. For many organisms, metamorphosis can be very attenuated or involves only a few tissues that undergo a functional or morphological transformation, especially in those organisms where many larval structures persist through adult life or where no change in habitat takes place as in fishes. In cases of extremely partial metamorphosis, it has not been possible to identify the regulatory hormonal signal.

Perhaps the most interesting diversity encountered with metamorphosis is among amphibia, ranging from a total absence of metamorphic change to partial to total transformation. Fig. 5.11 gives a few examples of amphibia that undergo partial or no metamorphosis. The latter are classified as neotenous, i.e., animals that do not undergo metamorphosis and exhibit larval biochemical and anatomical characteristics throughout their lifetime (gills, larval hemoglobin and visual pigment, etc.). They are thus capable of reproducing in the larval form. Neotenic amphibia can be divided into two classes: those that are obligatorily neotenous and those that are facultatively neotenous. Axolotls *(Ambystoma)* and some newts and salamanders (i.e., *Diemictylus*) are facultatively neotenic, a state which arises out of an inability of their thyroid gland to produce the hormone due to a defect in their pituitary to make or secrete thyrotropic hormone. Exogenous TH or TSH will under certain conditions, induce partial metamorphosis such as loss of tail fin and gills, keratinization of skin, appearance of lungs, etc. The Proteida family of amphibia comprise the obligatory neotenic species such as the American mudpuppy *(Necturus macalosus)* and the European *Proteus anguinus.*[51,52,54] Most interestingly, these species have a functional thyroid gland such that TH can be detected in their blood and administration of large doses of TH fails to induce any signs of metamorphosis.[2,54,55] What is the cause of this refractoriness to respond to thyroid hormone? Although a definitive answer is unlikely at the present, some recent work on comparisons of the structure and expression of thyroid hormone receptor (TR) genes in facultative and obligatory neotenic amphibia has given useful clues to the important role of TR gene expression in the metamorphic process. This aspect will be considered in detail in the next chapter (chapter 6).

Is Amphibian Metamorphosis Relevant to Mammalian Postembryonic Development?

In considering metamorphosis as a model for postembryonic development, an important question arises as to what extent it is relevant to mammalian postembryonic development. Because of the slow and continuous nature of the developmental changes occurring in late mammalian embryo or fetus, in contrast to the abrupt changes during metamorphosis, it is difficult to draw close parallels. Furthermore, unlike in metamorphosis which is induced and regulated by one or two hormonal signals, mammalian postembryonic development is complicated by the fetus being subjected to a number of overlapping and discontinuous maternal and fetal hormonal and growth factor signals, some not yet even identified. Nevertheless,

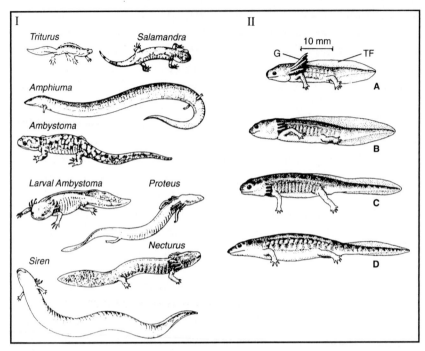

Fig. 5.11. I) Examples of some amphibians that undergo partial or no metamorphosis (neoteny). *Necturus* and *Proteus* are obligatorily neotenic whereas the other species undergo only partial metamorphosis or are facultatively neotenic, i.e., when exogenous thyroid hormone is provided. II) An example of facultative neoteny. Shown here is the axolotl or *Ambystoma* undergoing partial metamorphosis from the aquatic larval form (A) through intermediate stages (B,C) to the "adult" or terrestrial form (D) in response to endogenous or exogenous thyroid hormone. Note the total resorption of the external gills (G) and substantial loss of the tail fin (TF). From Tata JR. Metamorphosis. Burlington, NC: Carolina Biological Supply Co, 1983.

there are some remarkable similarities with many of the developmental changes induced by thyroid hormone during amphibian metamorphosis (Table 5.2) and those seen in the developing fetus or during the perinatal period in mammals (Table 5.4). For example, the activation during metamorphosis of the adult hemoglobin and the silencing of larval hemoglobin genes (see Fig. 5.8) has a parallel in the embryonic to fetal hemoglobin transition in man. Similarly, there is a parallel situation in the switch from the expression of α-fetoprotein to albumin in mammalian liver. Skin keratinization, bone and limb development and induction of urea cycle enzymes are other examples of this resemblance.

Making comparisons between free-living amphibian embryo and larvae and intra-uterine developing mammalian fetus is vitiated by the fact that it is impossible to identify with certainty the source and nature of the mammalian hormonal and other signals initiating and sustaining postembryonic development. Therefore, one is left to indirectly correlate the availability of a given hormone to the fetus, either from the fetal or maternal endocrine system, with the changes taking place in a given developing tissue. One experimental approach to partially obviate this difficulty is to block the supply of a given hormone or growth factor to fetal tissues. A most striking parallelism is particularly seen for the developing mammalian brain and bones, and the availability of thyroid hormones to the fetus and for behavioral imprinting and sex steroid hormones in fetal brain (see refs. 15, 56-59). Many of the structural and functional changes taking place in the metamorphosing frog tadpole brain[60] can also be seen during perinatal mammalian development, where thyroid hormones also play an important role.[61-64]

Perhaps the best documented illustration of the importance of hormones during human fetal development is that of cretinism caused by a deficiency of thyroid hormone to the fetus at late stages of pregnancy. The example of severe cretinism shown in Fig. 5.12 is caused by an almost total lack or even an absence of thyroid hormone normally provided by the mother to the fetus during gestation, as well as a failure of the fetal thyroid gland to function. It is well known that the consequence of this hormonal deficiency after birth is stunted growth, abnormal bone development and most characteristically, severe mental retardation accompanied by neurological lesions often resulting in deaf-mutism. If thyroid hormone replacement therapy is not initiated soon after birth, the lesions of cretinism or severe hypothyroidism are irreversible (see refs. 15,57,64). Many experimental studies on brain development in mammalian fetuses and newborns have highlighted the important influence exerted by thyroid hormones on the morphological and functional maturation of various neural cell-types. Differentiation of astrocytes, glial cells and various cerebellar and hippocampal cell-types is particularly sensitive to thyroid hormone.[62,63,65-68] Furthermore, with the availability of sensitive radioimmunoassays for thyroid hormones and TSH, it has been possible to temporally correlate the critical period for the requirements of these hormones and the establishment of important neurological and growth functions during postembryonic development.

As shown in the idealized representation of Fig. 5.13, the period of a few weeks before birth is marked by an abrupt appearance of thyroid hormones in human fetal blood. Depending on the developmental stage, the mammalian fetus and neonate derives its supply of thyroid hormone from both the maternal and its own thyroid gland. Only a narrow time window is available for the developmental actions of TH on the brain and already mentioned above, low levels of the hormone during this period will lead to irreversible damage and abnormal functions, such as cretinism, mental retardation and growth arrest.[61,64,66,69] A large body of work on rodents over the last 50 years has established that an experimental deficiency of TH during this critical perinatal period will soon lead to severe impairment of sensory functions and mental ability. Thus, it is significant that the rapid rise in TH levels in human fetal blood towards the end of gestation, shown in Fig. 5.13, is followed by an increased proliferation of glial and neuronal cells during the perinatal period. This in turn is accompanied by the acquisition of several cerebral functions as exemplified by sensory processes in Fig. 5.13. Significantly, TH is known to exert a strong influence on olfactory and optical patterning and function during amphibian metamorphosis.[60,70] It should be realized that the curves in Fig. 5.13 represent a correlation and not a cause-effect relationship in the developing mammalian fetus. For this reason, the recent demonstration by Forrest and his colleagues[71] that mice in which the TRβ gene had been "knocked out" by homologous recombination caused a profound loss of auditory function in newborn animals, even if they were supplied with exogenous thyroid hormone. It is also known from organ culture experiments with fetal rodent brain tissues that not only is thyroid hormone required for the increasingly complex neural and mental activities, but it is also necessary for reorganization of some parts of the developing vertebrate central nervous system. In invertebrates, ecdysone and juvenile hormone also play a major role in the structural and functional development of the insect CNS during metamorphosis.[7,72,73]

Although not directly relevant to metamorphosis, another example of the important role of hormones in mammalian brain development during fetal and perinatal life is that of sex steroids in establishing sexual dimorphism. Related to this hormonally regulated process of postembryonic development is sexual imprinting, whereby critical functions determining sexual and behavioral characteristics of adult life are irreversibly established.[58,59,74] An extreme example of this type of hormonal regulation is encountered in non-

Fig. 5.12. Photograph showing a typical cretin with stunted growth, mental retardation and deaf-mutism due to a severe deficiency of thyroid hormone during fetal life, in a region of South Asia. A normal adult relative is next to the affected individual.

mammalian vertebrates (birds, reptiles, amphibia) where a shift in the balance between androgen and estrogen at a well-defined time in embryonic or postembryonic development will alter or reverse phenotypic (in contrast to genetic) sex later in development. Alterations in the amounts of these sex hormones reaching certain regions of the brain of the developing mammalian fetus or early neonate will have far-reaching consequences in sexual and behavioral characteristics throughout later life.[58,59] These sensitive sites are characterized by a spatially well-defined presence of specific androgen and estrogen receptors, which are expressed early in development, as illustrated for estrogen receptor localization in Fig. 5.14. Again, an important feature of this imprinting phenomenon is the narrow time window during development only during which the specific target sites in the CNS for sex steroids remain highly sensitive to these hormones. A failure of the correct timing of sexual imprinting can result in abnormal reproductive activity and accessory sexual tissue formation, sterility and excessively aggressive or submissive behavior.[58] Interestingly, the estrogen receptor expressed during development in highly specific regions of the brain, such as those that

Table 5.4. Examples of mammalian postembryonic developmental expression of adult genes and processes in fetal tissues

Process	Genes and tissues
Gene switching	α-Fetoprotein to albumin; fetal to adult hemoglobin; immunoglobulin changes
Morphogenesis	Maturation and terminal differentiation of limbs, lungs, bones
Neural differentiation	Extensive neuronal cell turnover, acquisition of new functional and behavioral characteristics
Tissue restructuring	Keratinization of epidermis, connective tissue, remodeling of gut
Hormonogenesis	Activation of hormone producing genes in endocrine tissues
Induction of new functions	Urea excretion, new cell adhesion molecules
Cell death	Removal of tissues or organs by induction of lytic enzymes
Sexual differentiation	Activation of genes for sex determination, differentiation of accessory sexual tissues

determine in early avian development the activation of singing and reproductive behavior in adulthood.[75,76]

As noted earlier (section 5.1), prolactin and juvenile hormone block all the processes of metamorphosis induced by thyroid hormone and ecdysteroids in amphibia and insects, respectively. Virtually nothing is known about the possible juvenilizing action of prolactin in mammalian postembryonic development. However, an important hormone to which the developing fetus is exposed is placental lactogen, the product of a prolactin-like gene expressed in the placenta.[15,77] How are the kinetics of placental lactogen secretion regulated, what is the receptor for this hormone in the developing fetus, what are its biochemical and physiological actions on fetal tissues? These questions have to be addressed if one is to understand the major processes underlying mammalian fetal development to the same extent that hormonal regulation of amphibian and insect postembryonic development is understood. It would also be particularly relevant to determine whether placental lactogen can delay or counteract the action of thyroid hormone in mammalian postembryonic development. Resolution of this intriguing question will

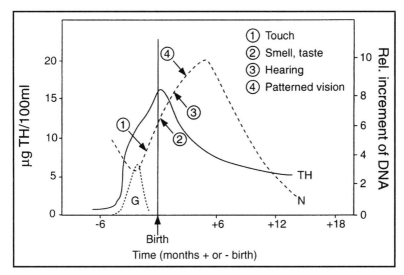

Fig. 5.13. Idealized curves showing the close association between the appearance of thyroid hormone (TH) in human fetal plasma, the proliferation of neuronal (N) and glial (G) cells in the brain, and the onset of acquisition of sensory perceptions (arrows) during the perinatal period and after birth. The proliferation of neural cells is represented as relative increments of DNA at different times during development. Deficiency of thyroid hormone during the perinatal period results in impairment of the development of these functions.

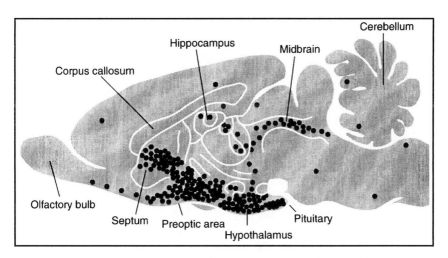

Fig. 5.14. Sexual differentiation of brain as depicted by the sites of localization of estrogen receptors in the rat. During a narrow time window around the time of birth, estrogen receptors, depicted as black dots, are expressed in well-defined regions of the brain which are concerned with sex-determined patterns of development, behavior and reproductive activity. Thyroid hormone also facilitates the imprinting by sex hormones. From Tata JR. Brain development and molecular genetics. In Dolentium Hominum: The Human Mind. Proceedings of the Fifth International Conference organised by the Pontifical Council, Vatican City, 1991:28-36.

considerably enhance our knowledge of what determines the timing of the acquisition of some of the important features of the adult phenotype in late fetal development and around the time of birth.

References

1. Etkin W, Gilbert LI, eds. Metamorphosis—A Problem in Developmental Biology. New York: Appleton-Century-Crofts, 1968.
2. Gilbert LI, Frieden E, eds. Metamorphosis: A Problem in Developmental Biology. New York: Plenum Press, 1981.
3. Gilbert LI, Tata JR, Atkinson BG, eds. Metamorphosis. Postembryonic Reprogramming of Gene Expression in Amphibian and Insect Cells. San Diego: Academic Press, 1996.
4. Gilbert LI, Rybczynski R, Tobe S. Endocrine cascade in insect metamorphosis. In: Gilbert LI, Tata JR, Atkinson BG, eds. Metamorphosis. Postembryonic Reprogramming of Gene Expression in Amphibian and Insect Cells. San Diego: Academic Press, 1996:60-107.
5. Russell S, Ashburner M. Ecdysone-regulated chromosome puffing in *Drosophila melanogaster*. In: Gilbert LI, Tata JR, Atkinson BG, eds. Metamorphosis. Postembryonic Reprogramming of Gene Expression in Amphibian and Insect Cells. San Diego: Academic Press, 1996: 109-44.
6. Wyatt GR. Insect hormones. In: Litwack G, ed. Biochemical Actions of Hormones. New York: Academic Press, II:1971; 386-490.
7. Riddiford LM, Truman JW. Biochemistry of insect hormones and insect growth regulators. In: Rockstein M, ed. Biochemistry of Insects. New York: Academic Press, 1978:307-56.
8. Balls M, Bownes M. Metamorphosis. Oxford, Clarendon Press; 1985.
9. Thummel CS. Flies on steroids—*Drosophila* metamorphosis provides insights into the molecular mechanisms of steroid hormone action. Trends Gen 1996; 12:306-10.
10. Riddiford LM. Juvenile hormone control of epidermal commitment in vivo and in vitro. In Gilbert LI, ed. The Juvenile Hormones. New York: Plenum Press, 1976.
11. Riddiford LM. Molecular aspects of juvenile hormone action in insect metamorphosis. In: Gilbert LI, Tata JR, Atkinson BG, eds. Metamorphosis. Postembryonic Reprogramming of Gene Expression in Amphibian and Insect Cells. San Diego: Academic Press, 1996:223-251.
12. Gudernatsch JF. Feeding experiments on tadpoles. Arch Entwicklungsmech. Organ 1912; 35:457-83.
13. Kaltenbach JC. Nature of hormone action in amphibian metamorphosis. In: Etkin W, Gilbert LI, eds. Metamorphosis. A Problem in Developmental Biology. New York: Appleton-Century-Crofts, 1968: 399-441.
14. Kaltenbach JC. Endocrinology of amphibian metamorphosis. In: Gilbert LI, Tata JR, Atkinson BG, eds. Metamorphosis. Postembryonic Reprogramming of Gene Expression in Amphibian and Insect Cells. San Diego: Academic Press, 1996:403-31.

15. Baulieu E-E, Kelly PA. Hormones. From Molecules to Disease. Paris: Hermann, 1990.
16. Kikuyama S, Kawamura K, Tanaka S et al. Aspects of amphibian metamorphosis: Hormonal control. Int Rev Cytol 1993; 145:105-48.
17. Wright ML, Cykowski LJ, Mayrand SM et al. Influence of melatonin on the rate of *Rana pipiens* tadpole metamorphosis in vivo and regression of thyroxine-treated tail tips in vitro. Develop Growth Differ 1991; 33:243-9.
18. White BA, Nicoll CS. Hormonal control of amphibian metamorphosis. In: Gilbert LI, Frieden E, eds. Metamorphosis. A Problem in Developmental Biology, New York: Plenum Press, 1981:363-96.
19. Tata JR. Gene expression during metamorphosis: An ideal model for post-embryonic development. BioEssays 1993; 15:239-48.
20. Tata JR. Hormonal interplay and thyroid hormone receptor expression during amphibian metamorphosis. In: Gilbert LI, Tata JR, Atkinson BG, eds, Metamorphosis. Postembryonic Reprogramming of Gene Expression in Amphibian and Insect Cells. San Diego: Academic Press, 1996:465-503.
21. Tata JR. Hormonal signaling and amphibian metamorphosis. Adv Dev Biol 1997; 5:237-74.
22. Tata JR, Kawahara A, Baker BS. Prolactin inhibits both thyroid hormone-induced morphogenesis and cell death in cultured amphibian larval tissues. Dev Biol 1991; 146:72-80.
23. Leloup J, Buscaglia M. La triiodothyronine, hormone de la metamorphose des amphibiens. CR Acad Sci 1977; 284:2261-3.
24. Tata JR. Metamorphosis. Burlington NC: Carolina Biological Supply Co, 1983.
25. Wald G. Metamorphosis: An overview. In: Gilbert LI, Frieden E, eds. Metamorphosis. A Problem in Developmental Biology. New York, Plenum Press, 1981:1-39.
26. Tata JR. Requirement for RNA and protein synthesis for induced regression of the tadpole tail in organ culture. Dev Biol 1966; 13: 77-94.
27. Ashburner M. Induction of puffs in polytene chromosomes of in vitro cultured salivary glands of *Drosophila melanogaster* by ecdysone and ecdysone analogues. Nature New Biol 1971; 230:222-4.
28. Ashburner M. Patterns of puffing activity in the salivary gland chromosomes of *Drosophila*. VI. Induction by ecdysone in salivary glands of *Drosophila melanogaster* cultured in vitro. Chromosoma 1972; 38:255-81.
29. Weber R, ed. Biochemistry of amphibian metamorphosis. In: The Biochemistry of Animal Development. New York: Academic Press, 1967:227-301.
30. Frieden E. Biochemistry of amphibian metamorphosis. In: Etkin W, Gilbert LI, eds. Metamorphosis—A Problem in Developmental Biology. New York:Appleton-Century-Crofts, 1968:349-98.
31. Cohen PP. Biochemical differentiation during amphibian metamorphosis. Science 1970; 168:533-43.

32. Clever U. Von der ecdysonkonzentration abhangige Genaktivita-tsmuster in den Speicheldrusenchromosomen von *Chironomus tentans*. Dev Biol 1963; 6:73-98.

33. Atkinson BG, Helbing C, Chen Y. Reprogramming of genes expressed in amphibian liver during metamorphosis. In: Gilbert LI, Tata JR, Atkinson BG, eds. Metamorphosis. Postembryonic Reprogramming of Gene Expression in Amphibian and Insect Cells. San Diego: Academic Press, 1996:539-66.

34. Weber R, Geiser M, Muller P et al. The metamorphic switch in hemoglobin phenotype of *Xenopus laevis* involves erythroid cell replacement. Wilhelm Roux's Arch Dev 1989; 198:57-64.

35. Weber R. Switching of globin genes during anuran metamorphosis. In: Gilbert LI, Tata JR, Atkinson BG, eds. Metamorphosis. Postembryonic Reprogramming of Gene Expression in Amphibian and Insect Cells. San Diego: Academic Press, 1996:567-97.

36. Andres AJ, Thummel CS. Hormones, puffs and flies: the molecular control of metamorphosis by ecdysone. Trends Gen 1992; 8:132-8.

37. Shi Y-B. Thyroid hormone-regulated early and late genes during amphibian metamorphosis. In: Gilbert LI, Tata JR, Atkinson BG, eds. Metamorphosis. Postembryonic Reprogramming of Gene Expression in Amphibian and Insect Cells. San Diego: Academic Press, 1996:505-38.

38. Brown DD, Wang Z, Furlow JD et al. The thyroid hormone-induced tail resorption program during *Xenopus laevis* metamorphosis. Proc Natl Acad Sci USA 1996; 93:1924-29.

39. Tata JR. Early metamorphic competence of *Xenopus* larvae. Dev Biol 1968; 18:415-40.

40. Baker BS, Tata JR. Accumulation of proto-oncogene c-*erb*-A related transcripts during *Xenopus* development: association with early acquisition of response to thyroid hormone and estrogen. EMBO J 1990; 9:879-85.

41. Weber R. Tissue involution and lysosomal enzymes during anuran metamorphosis. In Dingle JT, Fell HB, eds. Lysosomes in Biology and Pathology, Vol. I. Amsterdam: North-Holland, 1969:437-61.

42. Atkinson BG. Biological basis of tissue regression and synthesis. In Gilbert LI, Frieden E, eds. Metamorphosis: A Problem in Developmental Biology. New York: Plenum Press, 1981:397-444.

43. Lockshin RA. Cell death in metamorphosis. In: Bowen ID, Lockshin RA, eds. Cell Death in Biology and Pathology. London: Chapman and Hall, 1981:79-121.

44. Yoshizato K. Biochemistry and cell biology of amphibian metamorphosis with a special emphasis on the mechanism of removal of larval organs. Int Rev Cytol 1989; 119:97-149.

45. Yoshizato K. Cell death and histolysis in amphibian tail during metamorphosis. In: Gilbert LI, Tata JR, Atkinson BG, eds. Metamorphosis. Postembryonic Reprogramming of Gene Expression in Amphibian and Insect Cells. San Diego: Academic Press, 1996:647-71.

46. Tata JR. Hormonal regulation of programmed cell death during amphibian metamorphosis. Biochem Cell Biol 1994; 72:581-8.

47. Beckingham Smith K, Tata JR. The hormonal control of amphibian metamorphosis. In: Graham C, Wareing PF, eds. Developmental Biology of Plants and Animals. Oxford: Blackwell, 1976:232-45.

48. Saunders JW Jr. Death in embryonic systems. Science 1966; 154: 604-12.

49. Just JJ, Kraus-Just J, Check DA. Survey of chordate metamorphosis. In: Gilbert LI, Frieden E. Metamorphosis: A Problem in Developmental Biology. New York: Plenum Press, 1981:265-326.

50. Highnam KC. A survey of invertebrate metamorphosis. In: Gilbert LI, Frieden E, eds. Metamorphosis: A Problem in Developmental Biology. New York: Plenum Press, 1981:43-73.

51. Bentley PJ. Comparative Vertebrate Endocrinology. Cambridge University Press, 1982.

52. Shaffer HB. Phylogenetics of model organisms: The laboratory axolotl, *Ambystoma mexicanum*. Syst Biol 1993; 42:508-22.

53. Sehnal F, Svacha P, Zrzavy J. Evolution of insect metamorphosis. In: Gilbert LI, Tata JR, Atkinson BG, eds. Metamorphosis. Postembryonic Reprogramming of Gene Expression in Amphibian and Insect Cells. San Diego: Academic Press, 1996:3-58.

54. Turner C, Bagnara JT. General Endocrinology. Philadelphia: WB Saunders Co., 1976.

55. Hedges S, Maxson LR. A molecular perspective on Lissamphibian phylogeny. Herpetological Monographs 1993; 7:27-42.

56. Tata JR. Brain development and molecular genetics. In: Dolentium Hominum: The Human Mind. Proceedings of the Fifth International Conference organised by the Pontifical Council, Vatican City, 1991:28-36.

57. Degroot LJ, Larsen PR, Hennemann G. The Thyroid and Its Diseases. New York: Churchill Livingstone, 1996.

58. Kandel ER, Schwarts JH, eds. Principles of Neural Science. New York: Elsevier, 1985.

59. McEwen BS, Coirini H, Danielson A et al. Steroid and thyroid hormones modulate a changing brain. J Ster Biochem Mol Biol 1991; 40:1-14.

60. Burd GD. Role of thyroxine in neural development of the olfactory system. In: ISOT X, Proc of the Tenth Int Symp on Olfaction Taste, GCS a.c. Press, Oslo, 1990.

61. Dussault JH, Walker P. Congenital Hypothyroidism. New York: McGraw-Hill, 1983.

62. Dussault JH, Ruel J. Thyroid hormones and brain development. Ann Rev Physiol 1987; 49:321-34.

63. Legrand J. Hormones thyroïdiennes et maturation du système nerveux central. J Phys Paris 1983; 78:603-52.

64. DeLong GR, Robbins J, Condliffe PG, eds. Iodine and the Brain. New York: Plenum Press, 1989.

65. Gould EG, Frankfurt M, Westlind-Danielsson A et al. Developing forebrain astrocytes are sensitive to thyroid hormone. Glia 1990; 3:283-92.

66. Clos J, Legrand C, Legrand J. Effects of thyroid state on the formation and early morphological development of Bergman glia in the developing cerebellum. Dev Neurosci 1980; 3:199-208.

67. Rami A, Rabie A, Patel AJ. Thyroid hormone and development of the rat hippocampus: morphological alterations in granule and pyramidal cells. Neurosci 1986; 19:1217-26.

68. Trentin AF, Rosenthal D, Moura Neto V. Thyroid hormone and conditioned medium effects on astroglial cells from hypothyroid and normal rat brain: factor secretion, cell differentiation and proliferation. J Neurosci Res 1995; 41:409-17.

69. Morreale De Escobar G, Ruiz-Marcos A, Escobar Del Rey F. In: Dussault JH, Walker P, eds. Congenital Hypothyroidism. New York: Marcel Dekker, 1983:85-126.

70. Hoskins SG. Metamorphosis of the amphibian eye. J Neurobiol 1990; 21:970-89.

71. Forrest D, Golarai G, Connor J et al. Genetic analysis of thyroid hormone receptors in development and disease. Recent Prog Hormone Res 1996; 51:1-22.

72. Weeks JC, Levine RB. Postembryonic neuronal plasticity and its hormonal control during insect metamorphosis. Annu Rev Neurosci 1990; 13:183-94.

73. Truman JW. Metamorphosis of the insect nervous system. In: Gilbert LI, Tata JR, Atkinson BG, eds. Metamorphosis. Postembryonic Reprogramming of Gene Expression in Amphibian and Insect Cells. San Diego: Academic Press, 1996:283-320.

74. Siegel GJ, Agranoff BW, Albers RW et al, eds. Basic Neurochemistry (4th Edn). New York: Raven Press, 1989.

75. Jacobs EC, Arnold AP, Campagnoni AT. Zebra finch estrogen receptor cDNA: cloning and mRNA expression. J Ster Biochem Molec Biol 1996; 59:135-45.

76. Balthazart J, Ball GF. Sexual differentiation of the brain and behavior in birds. Trends Endocrin Metab 1995; 6:21-9.

77. Southard JN, Talamantes F. Placental prolactin-like proteins in rodents: variations on a structural theme. Molec Cell Endocrinol 1991; 79:C133-C140.

CHAPTER 6

Nuclear Receptor Expression During Postembryonic Development

The extensive involvement of hormones in a wide range of growth and developmental processes has quite understandably focused attention on the nature and function of their receptors. Changes in receptor activity accompanying the various cellular functions regulated by hormones, and signaling molecules in general, are by and large not due to alteration in their structure or active sites, but due to variations in the amount of the functional receptor proteins. Many early studies had established that long-term exposure to a growth and developmental hormone often leads to a reduction in the concentration of its own receptor. As several growth and developmental processes such as the ontogenesis of the lactating mammary gland or maturation of the egg are dependent on multiple hormones, it became increasingly evident that one of the participating hormones in such a complex system could also modulate the concentration of the receptor of another hormone also involved in regulating the growth and development of the same target tissue. As nuclear receptors play a predominant role in determining the rate and nature of postembryonic development (chapters 4,5), the next two chapters deal with auto- and cross-regulation of these receptors. The discussion below deals largely with the regulation of expression of nuclear receptors by their own ligands, and is largely exemplified by metamorphosis and vitellogenesis as models for postembryonic development.

Receptor Downregulation as a Classical Concept in Endocrinology

To a large extent our current understanding of the dynamics of hormonal regulation is based on the concept of negative feedback

Hormonal Signaling and Postembryonic Development,
by Jamshed R. Tata. © 1998 Springer-Verlag and R. G. Landes Company.

inhibition. Once it became possible to identify and measure receptor activity, the notion of homeostasis in adult organisms was often explained in terms of a given hormone controlling the activity or concentration of its own receptor.[1-4] Numerous studies on adult organisms in vivo, organ cultures and cell lines have described this phenomenon of receptor autoregulation as one of down-regulation for both membrane and nuclear receptors. For example, as depicted in Fig. 6.1, hyperinsulinemia has been shown to rapidly reduce the number of insulin receptor or binding sites in target tissue membranes.[1,5,6] Similarly, excessive estrogen or progesterone down regulates the number of estrogen or progesterone receptors in adult uterus nuclei.[2,4] Conversely, a deficiency of hormone can generally lead to receptor upregulation. These phenomena of up- and down-regulation of its own receptor have been recorded for virtually every hormone in adult animal tissues and in permanent cell lines. It is when one examines hormonal interplay underlying physiological processes during hormonally regulated growth and development that receptor upregulation becomes really obvious.[7]

Autoinduction of Nuclear Receptors During Growth and Development

The phenomenon of auto-upregulation of receptors is particularly noticeable with nuclear receptors when studying the regulation by the cognate hormones of the expression of specific genes or gross tissue development. This process is largely illustrated by studies from the author's laboratory.[7-9]

Auto-Upregulation of ER During Egg Protein Gene Expression

Chapter 4 has emphasized the valuable contribution made in early studies on hormonal regulation of gene expression by the model system of the activation by estrogen of genes encoding egg white or coat proteins and vitellogenin (and other yolk proteins) in the oviduct and liver of oviparous vertebrates, respectively.[10-14] It was realized early on that the hormonal induction of transcription of ovalbumin and vitellogenin genes was to a substantial extent dependent on ongoing synthesis of some proteins whose nature remained unknown until a few years later.

In 1978, Westley and Knowland[15] had suggested a possible link between *Xenopus* estrogen receptor (ER) and vitellogenin (Vit or VTG) synthesis. To extend our studies on the de novo activation by estrogen (E_2) of transcription of Vit genes in the liver of male *Xeno-*

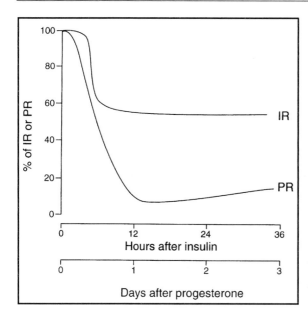

Fig. 6.1. Downregulation of a membrane and nuclear receptor by its own ligand. Idealized curves showing how insulin down regulates insulin receptor (IR) in human lymphocytes and progesterone reduces the concentration of progesterone receptor (PR) in adult rodent uterus. For further details see ref. 3.

pus, my laboratory made extensive use of primary cultures of male and female *Xenopus* hepatocytes in order to gain a better insight on the early events ensuing the exposure of these cells to estrogen[16-18] (see also Fig. 4.6). In one such study,[18] a close coupling was observed between the numbers of tightly bound ER in nuclei and the de novo induction of vitellogenin mRNA following the addition of E_2 to primary cultures of male *Xenopus* hepatocytes (Fig. 6.2A). Within 12 hours of addition of E_2 to these cultures, there occurred a 10- to 20-fold increase in the number of functional ER, as determined by ligand binding assays. A similar autoinduction of ER and its mRNA was observed in Shapiro's laboratory in male *Xenopus* liver in vivo.[12] In the same cultured male *Xenopus* hepatocytes in which ER was autoinduced (Fig. 6.2A) dr the first 12 hours after the addition of estrogen, there was a tightly coupled onset of the transcriptional activation of the normally silent Vit genes. At later times, whereas Vit mRNA continued to accumulate at a rapid rate, the concentration of ER reached a steady-state level. In whole animals, the latter was maintained for a period of several weeks after a single administration of E_2, even though the transcription of Vit genes had ceased within 4-8 days, if no further hormone was given.[19] This persistence of a high level of ER may partly explain the "memory effect," i.e., the more rapid and higher rate of Vit and ovalbumin gene expression, in amphibian and avian liver and oviduct (see chapter 4, Fig. 4.4). In

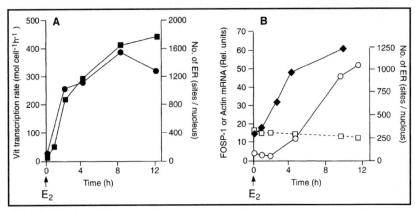

Fig. 6.2. A) Tight coupling between the concentration of nuclear estrogen receptor (ER), as measured by binding of estradiol-17β (E₂), and the absolute rate of vitellogenin (Vit) gene transcription. ER (■) and Vit transcription (●) were measured in the same batch of Xenopus liver nuclei as a function of time after the addition of E₂ to primary cultures of male Xenopus hepatocytes. For further details see ref. 18.
B) Correlation between the steady-state levels of Frog oviduct specific protein-1 (FOSP-1) mRNA (○) and nuclear ER (▼) in primary cultures of Xenopus oviduct cells at different times after the addition of estradiol. Actin mRNA (□) was also measured as a control for FOSP-1 mRNA. For further details see ref. 22.

the case of *Xenopus* hepatocyte cultures, it was also possible to demonstrate with cycloheximide and antiestrogens, that the absolute rate of transcription of Vit genes was a direct function of the accumulation of ER in the first 12 hours after the addition of E₂ (data not shown).[18] A similar co-induction of ER and Vit mRNA has also been observed for the response to E₂ of male trout hepatocytes in culture.[20]

The regulation of Vit gene expression in female avian and amphibian liver essentially follows the same pattern as that described above for male liver. On the other hand, in female oviparous vertebrates E₂ induces a different set of genes in the oviduct in concert with the activation of yolk protein genes in liver, the hormone therefore serving to coordinate the maturation of the egg (see Fig. 2.6). In *Xenopus* oviduct E₂ activates the expression of a set of genes encoding the egg coat or jelly-coat proteins, termed frog oviduct specific proteins or FOSP.[21-23] A pattern of coupled increase in nuclear ER and the accumulation of one of the FOSP genes (FOSP-1) similar to that seen for Vit gene expression in *Xenopus* hepatocytes in culture was also reproduced in oviduct cultures (Fig. 6.2B). The only difference between the two tissues was the kinetics of co-induction (compare Figs. 6.2A and B). Cycloheximide and antiestrogens also gave similar results with oviduct and liver cells in the autoinduction of

ER is tightly coupled to, and may even be necessary for the activation of the respective target tissues. As with steroid receptors in general,[2,4] no tissue-specific differences have yet been found which could explain the activation of different sets of genes with the same receptor in the liver and oviduct. More intensive exploration of tissue-specific factors determining which genes are selected for regulated expression and which are not in different target tissues of a given hormone would not only help in furthering our understanding of nuclear receptor function, but also of the role of multiple transcription factor interactions underlying tissue-specific gene expression in general (see chapter 4).

Autoinduction of TR Transcripts During Amphibian Metamorphosis

The expression of TRα and β genes in *Xenopus* tadpoles is under developmental control. Very small amounts of transcripts can be detected in unfertilized eggs and early embryos. A substantial increase, particularly of TRα mRNA, occurs at around stage 44, which quite significantly is when the *Xenopus* tadpole first exhibits competence to respond to exogenous thyroid hormone[24] (see chapter 5, Fig. 5.9). At this stage of development, several tissues which are programmed to undergo major changes later during metamorphosis show high concentrations of TR mRNA. In the examples of in situ hybridization patterns presented in Fig. 6.3, this is clearly seen in the brain, liver, small intestine and tail.[25] At later stages (around 55), high concentrations of the transcripts are also found in limb buds (data not shown). It is worth recalling that all these tissues undergo quite different changes during metamorphosis leading to the acquisition of the adult phenotype.

After stage 54 and until the completion of metamorphosis there is good correlation between the accumulation of TR transcripts and the circulating level of thyroid hormone in *Xenopus* tadpoles (see Figs. 5.2 and 6.4 and also refs. 7,8,26,27). As seen in Fig. 6.4, the relative amounts of TRα and β mRNAs vary according to the region of the tadpole, TRβ being more strongly expressed in the head region, and also according to the progression of metamorphosis after stage 54.

The correlation between the increase in TR mRNA and circulating thyroid hormone raised the question as to whether or not exogenous TH would precociously upregulate TR gene expression

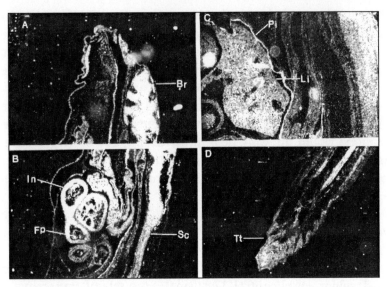

Fig. 6.3. In situ hybridization of TR mRNA in different regions of the early premetamorphic Xenopus tadpole. Saggital sections of A) the head, B) the middle, C) the gut and D) the distal part of the tail of stage 44 tadpole are presented. The bright zones of the dark-field images shown here reveals strong localization of TR transcripts in brain (Br), spinal cord (Sc), liver (Li), tail tip (Tt). Pigmented layers (Pi) and food particles (Fp) also appear as bright zones. For details see ref. 25.

in pre-stage 54 tadpoles. Several experimental studies from the laboratories of Brown, Shi and Tata, based on Northern blotting, RNase protection assays and in situ hybridization of RNA from various *Xenopus* tadpole tissues all established that indeed T_3 causes a substantial autoinduction of TR mRNA.[7,9,25,27–29] A similar upregulation of TR mRNA has been observed in *Rana catesbeiana* tadpoles.[30] Some results obtained with in situ hybridization are illustrated in Fig. 6.5. Enhanced TR signals can be easily discerned in brain, intestine, and liver of pre-metamorphic (stage 52) *Xenopus* tadpoles exposed to low doses of T_3 for 4 days. A similar increased accumulation of TR mRNA is also visible in stage 53 tadpoles hind limb buds following T_3 treatment for 5 days. Note that in stage 53 tadpoles, 5 days of T_3 treatment has produced considerable growth of the hind limb bud accompanying the enhanced accumulation of TR mRNA (compare Fig. 6.5 g and h). The same phenomenon of autoinduction of TR mRNA is observed for *Xenopus* tadpole tails in vivo and in organ culture (see chapter 8). All these tissues exhibit the initiation of different programs of structural and biochemical modifications, de novo morphogenesis and cell death during both normal and hormone-induced metamorphosis (see ref. 31).

It is worth considering some important features of TR autoinduction to assess its significance in the action of TH in regulating metamorphosis:

1) The extent of autoinduction of TR is dependent on the developmental stage of the tadpole. T_3 is ineffective at the very early developmental stages, i.e., prior to stage 44 (see chapter 5). However, as shown in Fig. 6.6, 3 days of exposure to low doses of T_3 of stage 48 tadpoles provokes a strong induction of TRβ mRNA. This inducibility increases with development, the highest sensitivity being reached as metamorphosis progresses towards its climax. At this point it is worth recalling that the albumin gene, which is a major target for TH during natural metamorphosis, is precociously inducible by T_3 added to tadpoles at stages 53 to 55 (see Fig. 5.10). Thus the albumin gene is only activated by exogenous T_3 when the basal levels of TRα and β mRNA (and presumably also of the receptors) are already beginning to be elevated (see Fig. 6.4). The magnitude of hormonal response increases after that stage until the mid-metamorphic and climax stages of 58-61 of *Xenopus* tadpoles. It correlates well with the increase in sensitivity to T_3 during metamorphosis.

2) The magnitude and kinetics of TR mRNA upregulation varies from tissue to tissue, cell type to cell type and in different regions of the tadpole (see Figs. 6.4, 6.5).

3) Autoinduction is marked more for TRβ mRNA than for TRα. In all organisms, including amphibia, TRα transcripts are about 10-20 times more abundant than those of TRβ. In pre-metamorphic tadpoles, TRβ mRNA is difficult to detect by Northern blotting, which has led some investigators to suggest that TRα is induced for the first time by the interaction between TRα and TH when the secretion of endogenous hormones is initiated at the onset of metamorphosis.[27] That this is unlikely is suggested by immunocytochemical detection of TR protein described below.

As regards kinetics of TR autoinduction, the relative amount of TRα mRNA in whole stage 52 *Xenopus* tadpoles is increased 2- to 4-fold 48 hours after T_3, whereas TRβ transcripts increased 20- to 50-fold (Fig. 6.7A). What is particularly important is that an

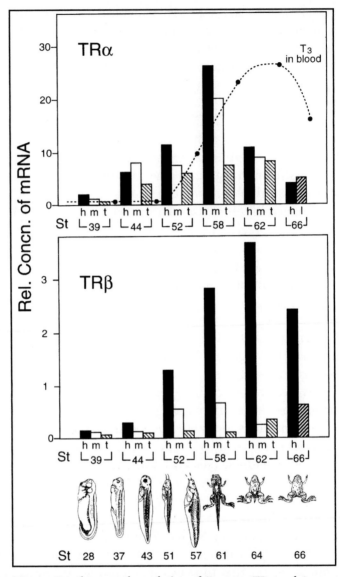

Fig. 6.4. Developmental regulation of Xenopus TRα and β gene expression during metamorphosis. RNA was extracted from head (h), middle (m) and tail (t) regions of tadpoles at different stages (St) before (39, 44, 52), during (58, 62) and after (66) natural metamorphosis. Broken line denotes the levels of circulating T_3 during this developmental period (see ref. 8). TR mRNAα accumulates in all tissues, and at all stages, to higher levels than does TRβ mRNA; note the 10-fold difference in scale for their relative amounts. Other details in ref. 25.

Table 6.1. Time that elapses before activation of different genes in premetamorphic Xenopus tadpoles induced to metamorphose precociously with T_3

Gene	Time Required (h)
TRβ	4
TRα	8
Albumin	40
Stromelysin 3	48
L-Arginase	70
63 kDa keratin	100

The values are based on the latent period preceding the de novo appearance or a 10% increase in amount of each mRNA in stage 52 Xenopus tadpoles treated with 10^{-9} M T_3.

upregulation of TRβ mRNA can be seen as early as 4 hours after the exposure of these pre-metamorphic tadpoles to exogenous T_3. As can be seen in Table 6.1, this is among the most rapid biochemical responses of *Xenopus* tadpoles. Furthermore, TRβ is a direct-response gene which, together with the rapid response, raises the possibility that upregulation of TR is a requirement for gene programming during hormonal induction of metamorphosis.

Autoinduction of TRα and β mRNAs can also be reproduced in permanent, non-transformed, *Xenopus* cell lines. Cultures of XTC-2 and XL-177 cells are particularly responsive to T_3.[32-35] Remarkably, the kinetics of upregulation of the two receptor transcripts in these cell lines were very similar to those induced in whole tadpoles, as illustrated in Fig. 6.7B for XTC-2 cells (compare Figs. 6.7A and B). Using these cultured cells, it was possible to show that the autoinduction TR mRNA is accompanied by that of functional TR protein by transfecting them with a DNA construct comprising bacterial chloramphenicol acetyl transferase (CAT) reporter fused to the promoter of *Xenopus* albumin gene (Fig. 6.8).[8] The latter gene is a well-known target for T_3 during amphibian metamorphosis[36-39] (see Tables 5.2, 6.1). When such transfected cells were exposed to T_3 under conditions in which the hormone induces TRα and β mRNAs, there was a significant increase in CAT activity after its transcription from the albumin promoter (Fig. 6.8). This finding strongly suggests that the autoinduction of TR mRNA leads to the production of more functional TR protein. It also raises the possibility that TR has to be raised to a higher concentration than is required for the

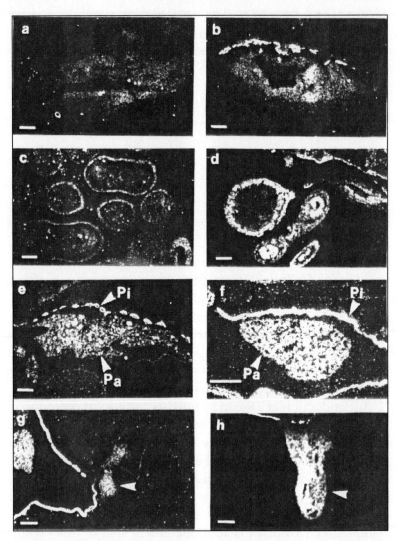

Fig. 6.5. Upregulation by T_3 of TR mRNA in brain (a,b), small intestine (c.d), liver (e,f) and hind limb buds (g,h) of premetamorphic (stage 52/53) Xenopus tadpoles. The illustrations are dark-field imaging of localization by in situ hybridization with antisense Xenopus TR cRNA (see Fig. 6.3). Saggital sections were cut from different tissues of control (a,c,e,g) or treated (10^{-9} M T_3 for 4 days; b,d,f, or 5 days for h) tadpoles. Sense probe gave virtually no signal and those images are therefore not shown. Arrows in g and h show hind limb buds. Arrows in e and f indicate parenchymal liver cells (Pa) and an artefact produced by pigmentation surrounding these cells (Pi). Bars are 100μm. For more detail see ref. 29.

Table 6.2. *Metamorphic response, endogenous TH production and expression and autoinduction of TRα and β genes in facultative and obligatory neotenic amphibia, compared with naturally metamorphosing Xenopus*

Species	Metamorphosis	Endogenous TH Produced	TR Genes Present	Expressed	Autoinduced
Xenopus	Spontaneous	Yes	Yes	Yes	Yes
Ambystoma	Facultatively neotenic	No	Yes	Yes	Yes
Necturus	Obligatorily neotenic	Yes	Yes	Only TRα	No

activation of its gene in order to induce one of the metamorphosis-specific target genes (see below).

Comparative studies on the response to exogenous thyroid hormones of facultative (e.g., axolotl) and obligatory (e.g., *Necturus*) neotenic amphibia (see Fig. 5.11) also indirectly point to the strong association between upregulation of TR and metamorphosis. As shown in Table 6.2, administration of T$_3$ to larval axolotls, whose thyroid glands are non-functional and will therefore not undergo metamorphosis normally, can be induced to undergo metamorphosis artificially as determined by several morphological and biochemical criteria (see chapter 5). In contrast, *Necturus*, which has a functional thyroid gland, does not respond to even high doses of exogenous TH. A possible explanation for the divergent responses of facultative and obligatory neotenic amphibia is offered by experiments in which autoinduction of TR genes by T$_3$ was determined[8,28] (see Table 6.2). Low levels of TR mRNAs were detected in axolotl (*Ambystoma mexicanum* and *Ambystoma tigrinum*) larvae, but not in those of *Necturus macalosus*. Treatment with T$_3$ of *Ambystoma* larvae upregulated TR mRNA, but not in tissues of *Necturus*. The autoinduction of axolotl TR genes was accompanied by metamorphosis, as externally visible by such indices as regression of gills and tail fin and keratinization of the skin. Interestingly, in a recent study[40] on the evolution of TR genes in amphibia, based on PCR amplification analysis, both species of axolotls as well as *Necturus* had well conserved TR α and β genes. Reverse transcriptase polymerase chain reaction (RT-PCR) analysis revealed the presence of TRα but not TRβ transcripts in some tissues of *Necturus*. As TRβ transcripts could be detected in *Ambystoma* tissues, this observation indirectly indicates that TRβ gene expression is essential for metamorphosis.

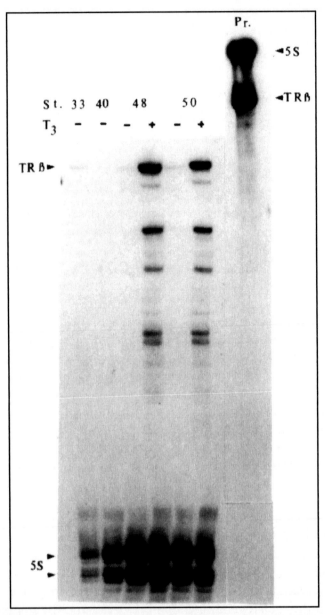

Fig. 6.6. Developmental stage-dependent early acquisition of metamorphic competence by Xenopus tadpoles to respond to thyroid hormone, as revealed by the activation of TRβ gene. Stage 33 embryos and tadpoles at stages 40, 48 and 50 were exposed to 2 nM T₃ for 3 days before RNA from whole larvae was extracted and the concentration of TRβ mRNA determined by RNase protection assay. The major TRβ and Xenopus 5S RNA (used as a control) bands are shown. Note that at the end of 3 days of treatment with T₃, the untreated larvae would have reached stages 38, 43, 50 and 51/52. Pr = probe.

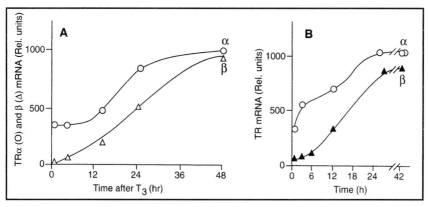

Fig. 6.7. Kinetics of autoinduction of TRα and β mRNA in whole stage 50/51 pre-metamorphic Xenopus tadpoles (A) and cultured *Xenopus* XTC-2 cells (B). Tadpoles and cell cultures were treated with 2 x 10⁻⁹ M T₃ for different periods of time shown, after which the relative concentration of TRα and β mRNAs were determined by RNase protection assay. For experimental details see refs. 25 and 32.

Autoinduction of TR Protein During Amphibian Metamorphosis

The facts that TH can rapidly upregulate TR transcription in the presence of inhibitors of protein synthesis,[27] and that early *Xenopus* tadpole stages respond to T_3,[24] indicates the presence of functional receptors well before the onset of natural metamorphosis. In contrast to the numerous studies on transcription of nuclear receptor genes, there are very few reports on tissue distribution or developmental and hormonal regulation of expression of receptor proteins.[4,41] More specifically, as regards amphibian metamorphosis, only two publications deal with TRα and β proteins during *Xenopus* development and metamorphosis. Using polyclonal antibodies to detect *Xenopus* TRs in embryo or tissue extracts, Eliceiri and Brown[42] were able to find TRα, but not TRβ, in unfertilized eggs, embryos and all stages of tadpoles before and during metamorphosis. These investigators could only detect TRβ at stages when endogenous TH was secreted, i.e., at the onset of natural metamorphosis, or if exogenous T_3 was administered to pre-metamorphic tadpoles. The amount of two proteins was found to roughly increase as a function of the enhanced accumulation of their respective mRNAs, which led them to suggest that TRβ was induced by the liganded TRα. However, when Fairclough and Tata[43] used specific monoclonal antibodies to detect the two receptor isoforms immunocytochemically, it was possible to demonstrate the presence of both TRα and β proteins in the nuclei of all tissues examined from tadpoles before and

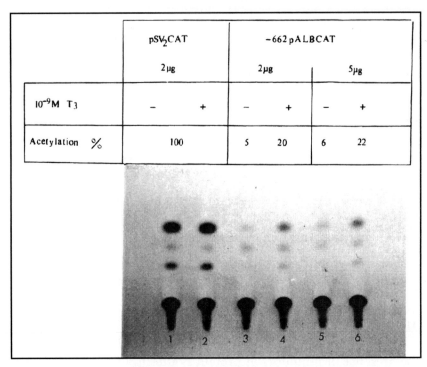

| | pSV$_2$CAT | −662 pALBCAT | |
	2 µg	2 µg	5 µg
10^{-9}M T$_3$	− +	− +	− +
Acetylation %	100	5 20	6 22

Fig. 6.8. Autoinduction of TR mRNA is accompanied by the activation of Xenopus albumin gene by T$_3$ in XTC-2 cells. XTC-2 cells were transfected with pSV$_2$ CAT as a control (lanes 1 and 2) or plasmid -662pALB-CAT comprising 662 nt of Xenopus albumin promoter incorporating a TRE and linked with a CAT reporter gene (lanes 3-6). Half the cells were untreated with T$_3$ (lanes 1,3,5) and the other half treated with 10^{-9} M T$_3$ (lanes 2,4,6). 24 hours later the acetylation was assessed by autoradiography, as shown, and quantified by liquid scintillation counting (not shown). Other details in ref. 8.

during natural metamorphosis. Administration of T$_3$ nevertheless augmented the amount of both proteins, as shown in Fig. 6.9, for TRβ in tadpole liver and small intestine. The first tissue undergoes extensive gene switching and functional maturation, whereas the latter loses about 90% of its cells followed by rapid morphogenesis. It was not possible to show immunocytochemically any equivalence between TR mRNA and protein content. In this context it is worth recalling that Lechan et al[44] found in the rat brain a lack of correlation between the variations of TRβ2 mRNA and proteins.

What causes the apparent discrepancy in the results obtained with polyclonal and monoclonal antibodies for *Xenopus* TRs in the above two studies is not clear. They could be due to several technical reasons. For example, in the studies from Brown's laboratory, TR

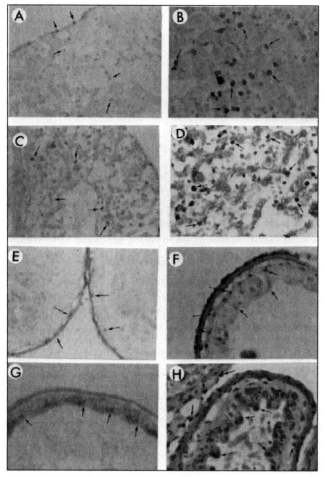

Fig. 6.9. Upregulation by T$_3$ of Xenopus TRβ protein in liver (A,B,C,D) and intestinal epithelium (E,F,G,H) of stage 52 (A,B,E,F) and 55 (C,D,G,H) tadpoles. The animals were either untreated (A,C,E,G) or exposed to 2.5 nM T$_3$ for 5 days (B,D,F,H) before sections were taken for immunocytochemical localization of TRβ. Arrows denote some of sites of TRβ accumulation. For details see ref. 43.

mRNA and protein were identified by Northern blotting and immunoprecipitation with polyclonal antibodies, whereas in our studies the comparisons were made with in situ hybridization and immunocytochemistry. Despite these differences, there is agreement between the two sets of data on one important point. From both studies it is clear that TH induces both the receptor mRNAs as well as proteins.

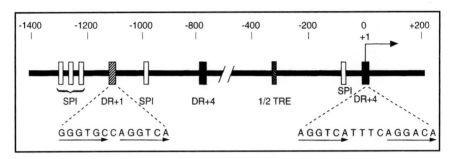

Fig. 6.10. Schematic illustration of the Xenopus TRβ gene promoter. The positions of the proximal and distal DR+4 TREs, a half TRE, the putative DR+1 9-cis-retinoic acid response element (RXRE) and five SP1 binding sites are shown. Details of the functionality of the DR+4 sites and other promoter sequences are in ref. 47.

What Is the Mechanism of Autoinduction of TR?

The most simple mechanism to explain the phenomenon of autoregulation of *Xenopus* TR would be a direct interaction between TR proteins and the promoters of the genes encoding them, providing the promoter contains one or more thyroid hormone responsive element (TRE). Since the expression of *Xenopus* TRβ gene is modulated by T_3 to a greater extent than that of the α gene (see above, Fig. 6.7)it is significant that the promoter of the β gene has been cloned and characterized.[45-47]

Two TREs have been identified in 1.6 kb of the upstream sequence of the *Xenopus* TRβ gene.[47] Both of these are of the direct hexanucleotide repeat separated by 4 nucleotides (DR+4) type. This TRE is the most common, functional response element in TH target genes (see chapters 3, 4). As summarized in Fig. 6.10, there is an imperfect distal DR+4 site at -793/-778 and a perfect proximal DR+4 at -5/+11 in the TRβ gene promoter. One DR+1 retinoid X response element (RXRE) at -1056/-1044 and five SP1 sites, but no TATA or CAAT boxes could be discerned. The presence of SP1 and interaction sites for other transcription factors, as well as the absence of TATA and CAAT boxes, are common features of promoters of many genes encoding nuclear receptors and some transcription factors.[41] The functional properties of the two DR+4 TREs was assessed by deletion analysis and cell transfection, in conjunction with the competitive binding of the TREs to recombinant *Xenopus* TRα and β proteins. Transfection of *Xenopus* XTC-2 and XL-2 cells, which express both TRα and β[32] (Fig. 6.7B) with CAT constructs of various TRβ promoter

Table 6.3. Auto-upregulation of nuclear receptors by some developmental hormones

Hormone and Receptor	Species and Tissue	Developmental Action
Thyroid hormone	All amphibian tissues	Metamorphosis
	Rodent astrocytes	Brain development
Estrogen	Avian and amphibian liver	Vitellogenesis
Ecdysone	All insect cells	Egg maturation
Retinoic acid	Rodent and avian brain, limbs, etc.	Metamorphosis
Androgen	Mammals; male accessory sexual tissues	Sexual maturation

fragments showed that both the proximal and distal TREs responded to T_3, the former element being more active (data not shown). Interestingly, overexpression of unliganded TRα and β in these *Xenopus* cell lines caused a substantial suppression of basal transcriptional activity. This transcriptional repression by unliganded TRs has been well characterized in mammalian systems (see chapter 4).[48-51] Under the conditions of transcriptional suppression, the addition of T_3 to *Xenopus* cells, co-transfected with the full-length TRβ promoter, produces up to 20-fold enhancement of CAT activity.[47]

Electrophoretic mobility shift assays (EMSA) performed with the TRβ promoter fragments and recombinant *Xenopus* TRα and β supported the findings of transcriptional activities. Both the proximal and distal TREs strongly interact with the receptors[47] but of the two TREs the proximal one was more potent. As shown in Fig. 6.11, a strong binding could only be observed when recombinant *Xenopus* RXRα, β or γ were added together with TRα or β. These EMSAs clearly demonstrate that TR-RXR heterodimers (in any combination), but not TR monomers or homodimers specifically interact with the DR+4 TREs of *Xenopus* TRβ gene promoter.[35,47] While these studies do not rule out the participation of different response elements or other accessory factors (see chapter 4), they strongly support the idea of a direct interaction between the thyroid hormone receptor and the promoter of its own gene as the most likely mechanism underlying *Xenopus* TR autoinduction.

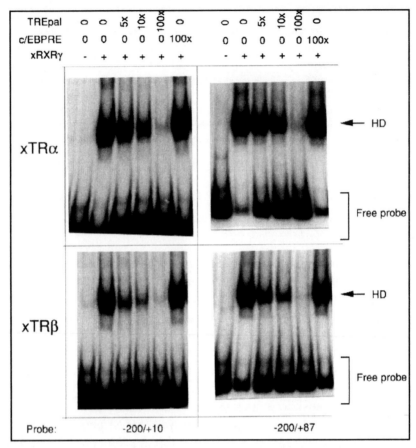

Fig. 6.11. Specific, competitive binding of the proximal DR+4 TRE (-5/+11) in the Xenopus TRβ gene promoter to recombinant Xenopus TRα and β (xTRα,β), in the presence or absence of recombinant Xenopus RXRγ (xRXRγ). The two ^{32}P-labeled DR+4 probes used and EMSA procedure are as described by Machuca et al.[47] TREpal is the consensus palindromic TRE used as competitor at 5-, 10- and 100-fold excess and c/EBPRE is a non-specific response element[76] used as a control at 100-fold excess over the probes. HD denotes the position of the TR-RXR heterodimer.

The Dominant Negative Receptor Approach

A direct approach based on gene 'knock-out,' as has been achieved in mice,[52] would provide strong evidence for the physiological significance of auto-upregulation of TR during *Xenopus* metamorphosis. But since homologous recombination has not yet been successfully achieved in amphibia, it became necessary to take other indirect approaches. One such is based on the use of dominant negative TRs.[35] Certain forms of truncated or mutant TRs characterize the human disease termed Generalized Thyroid Hormone Resistance (GTHR) syndrome.[53-55] Virtually all the mutations described involve deletion, frame-shift or addition in or around the ligand-binding domain of the β gene. Like its viral homolog v-*erb*A,[56] the mutant receptor fails to bind TH and will prevent the liganded normal wild-type receptor from interacting with TREs and activating transcription (see Fig. 6.12A). This dominant-negative activity renders the mutants useful tools in exploring various facets of TR functions.

Recently Ulisse et al[35] have exploited these naturally occurring and artificially generated human and *Xenopus* dominant-negative TRs to delve further into the mechanism of autoregulation of TR and its significance in T_3-induced metamorphosis. As shown in Fig. 6.12B, the human dominant-negative mutant TR and the synthetic *Xenopus* TRβ construct, homologous to a C-terminal human mutant receptor, abolished the ability of T_3 to upregulate endogenous or over-expressed wild-type *Xenopus* TRβ when these were transfected into *Xenopus* XTC-2 cells. The recombinant mutant TR proteins heterodimerized with *Xenopus* RXR to interact strongly with the DR+4 TREs present in *Xenopus* TRβ gene promoter (data not shown). These and other results are compatible with the mechanism proposed for dominant-negativity, whereby the mutant receptor unable to bind the ligand forms a complex with its wild-type heterodimeric partner and thus acts as a transcriptional repressor.[54] Ulisse et al[35] extended these experiments in *Xenopus* cell lines to the co-injection of dominant-negative and wild-type receptor constructs with TRβ TRE-CAT reporter into pre-metamorphic *Xenopus* tadpole tails. Culturing these tails in the presence of T_3 (see chapter 8) reproduced the repressor activity of dominant-negative mutants in vivo. This inhibition of upregulation of TRβ, which is one of the most rapid, direct responses to T_3, now allows one to further establish the importance of TR autoinduction for metamorphosis by extending these studies to the expression of target genes downstream from the receptor.

Dominant-negative nuclear receptors for other than thyroid hormone have also been described for other nuclear receptor-associated disorders in man. A large number of mutations of human androgen receptor, which is a X-chromosome-linked gene, have now been described in clinical cases and which are the cause of androgen insensitivity (see refs. 57,58 for reviews). The androgen insensitivity

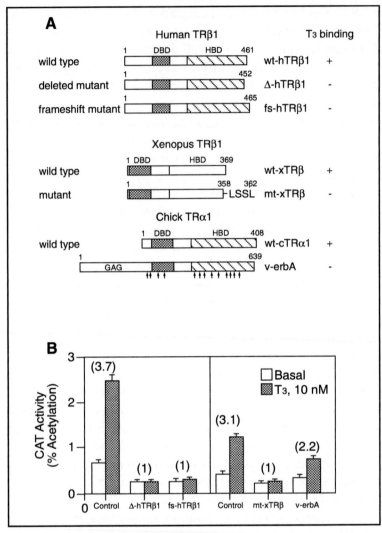

Fig. 6.12. A) Organization and ligand binding properties of wild-type and mutant human, Xenopus and chicken TR and the viral oncogene v-*erb*A. One of the mutant human TRβs (△-hTRβ1) has a deletion of the last 9 amino acids, while the other (fs-hTRβ1) is a frameshift mutation caused by a 7-nucleotide duplication. The artificially mutated Xenopus TRβ (mt-xTRβ) has the last 11 amino acids of the wild type receptor (wt-xTRβ) replaced by 4 amino acids. The v-*erb*A oncogene shows N- and C-terminal deletions along with several internal mutations (arrows) compared with the wild-type chicken TRα1 (wt-cTRα1). The T_3 binding properties are denoted as + (normal) and -no) binding. DBD and HBD denote DNA- and hormone-binding domains, respectively. The numbers refer to amino acid positions. B) Dominant-negative activity of mutant human and Xenopus TRβ and v-*erb*A in Xenopus XTC-2 cells. The cells were transfected with a wild-type Xenopus TRβ promoter with a CAT reporter (p[-1500/

syndrome covers a wide range of phenotypic abnormalities in male sexual development that are caused by accessory sexual and other tissue resistance to androgen action. Mutations in this receptor gene are more widespread than reported for other nuclear receptors and include missense mutations causing amino acid substitutions, deletions, mRNA splice site alterations and nonsense mutations. A dominant-negative nuclear receptor of considerable interest would be that of RXR because of its involvement as a heterodimeric partner of the subgroup of non-steroidal nuclear receptors (see chapter 3). Although RXR mutations associated with human diseases have not been identified, it is of some significance that a synthetic mouse RXRβ with a mutation in the DNA binding domain was found to inhibit the response of mouse embryonal carcinoma cells to retinoic acid.[59] Construction and exploitation of RXR mutants with defective ligand-binding and heterodimerization domains will prove to be particularly valuable tools in exploring the functional characteristic of the subgroup of nuclear receptors that require RXR for their activity.

Auto-Upregulation of Other Nuclear Receptors

Besides the examples of autoinduction of ER and TR, described above in detail, that of other nuclear receptors has also been documented, but in lesser detail. Some examples[7,8] of nuclear receptor auto-upregulation are presented in Table 6.3. Thus, all three isoforms of the mouse retinoic acid receptor (RARα,β,γ) have been shown to be induced by retinoic acid.[60,61] Interestingly, retinoic acid response elements (RAREs) have been identified in the promoters of RARβ and γ genes and that retinoic acid has been shown to operate through these response elements to upregulate the expression of its own receptors.[60,62] The three genes of the RAR family give rise to multiple sub-isoforms with different tissues exhibiting different patterns of their distribution. Chambon and his colleagues have suggested that this differential isoform distribution and the retinoic acid receptor autoinduction have an important role to play during mammalian embryonic and fetal development.[61,63,64] Among other examples of

---

+87]xTRβ-CAT) construct with or without the TR mutants Δ-hTRβ1, fs-hTRB1, mt-xTRβ or v-*erbA*, as depicted in (A). CAT activity was measured and the values in parenthesis are fold-stimulation of activity when the cells were incubated with T₃. Note that all three mutant TRβs, but not v-*erbA*, cause a reduction of basal CAT activity. For details see ref. 35.

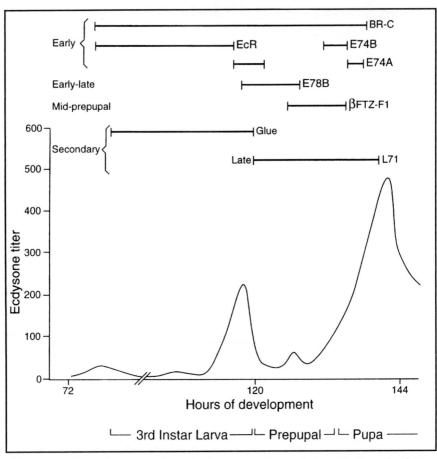

Fig. 6.13. Correlation between pulses of ecdysone secretion and the activation of ecdysone receptor (EcR) and some early and late developmental genes during metamorphosis of *Drosophila*. The curve at the bottom of the figure represents the high and low pulses of ecdysone during the 3rd instar larval, prepupal and pupal stages. The horizontal bars represent the detection of mRNAs encoded by EcR, early (BR-C, E74B, E74A), early-late (E78B), mid-prepupal (βFT2-F1) and secondary or late (Glue, L71) genes during metamorphosis. For a description of these genes and other details, see refs. 69 and 70.

autoregulation of nuclear receptors in mammals, are those of androgen receptor in male accessory sexual tissues[65,66] and of TRβ in the central nervous system.[67]

Thummel's group have carried out a detailed analysis of the upregulation of ecdysone receptor (EcR) in *Drosophila* at all stages of early- and postembryonic development.[68-70] Thanks to the advantages of genetic analysis in this organism, it has been possible for them to accurately establish the temporal coordination of ecdysone

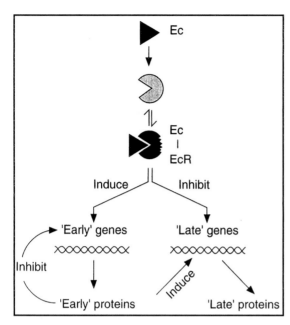

Fig. 6.14. Model proposed by Ashburner et al[72] to explain the dual effects of ecdysone (Ec) on the induction and de-induction of serial chromosomal puffs during metamorphosis in *Drosophila*.

It is proposed that the liganded ecdysteroid receptor (EcR) activates 'early' puff genes but represses 'late' genes. The products of 'early' genes would, in turn, induce the 'late' genes but inhibit the 'early' genes themselves, thus causing the puffs to disappear.

receptor expression with that of other regulatory genes before, during and after multiple bursts of metamorphic activity in *Drosophila*[70,71] The example given in Fig. 6.13 illustrates how the burst of EcR expression at the onset of larval metamorphosis anticipates that of other regulatory genes during metamorphosis. A characteristic feature of insect postembryonic development is the pattern of temporal coordination of both the accumulation and disappearance of products of different regulatory genes with the multiple pulses of ecdysone secretion during the sequential metamorphic responses considering the autoinduction of EcR. It is therefore appropriate to recall that more than 25 years ago Ashburner had proposed the possibility of autoinduction of ecdysone receptor, followed by its downregulation[72,73] (see Fig. 6.14). This model has stood the test of time and what is particularly important is that it was put forward well before the discovery of nuclear receptors and the advent of molecular techniques for their identification and quantification. The validity of the Ashburner model has been upheld by all the recent investigations on EcR gene expression. To conclude, it is clear from the above account that receptor auto-upregulation is a general phenomenon during postembryonic development and not just restricted to thyroid hormone receptors and amphibian metamorphosis.

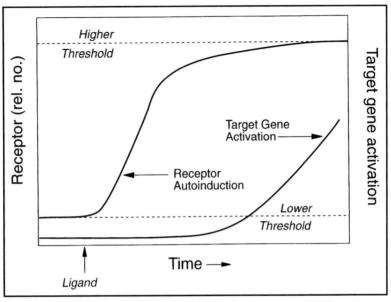

Fig. 6.15. A simplified dual-threshold model to explain the physiological significance of auto-upregulation of nuclear receptors in hormonally regulated developmental processes. According to this model a given receptor is constitutively expressed at a level too low to activate the target genes that specify a particular developmental function (lower threshold). Upon the release of the hormone the liganded receptor is capable of inducing its own receptor by virtue of a high affinity interaction with the promoter of the gene encoding the receptor. When the latter reaches a certain higher threshold level it becomes capable of activating the downstream target genes responsible for the phenotypic response.

Significance of Autoinduction of Nuclear Receptors

While a direct cause-effect relationship has not been established between autoinduction of nuclear receptors and the physiological action of the hormone in postembryonic development, there is sufficient evidence now to suggest that receptor upregulation is closely linked to the biological activity of its ligand. It is therefore appropriate to close this chapter by considering a general model which attempts to explain the significance of the phenomenon of receptor autoregulation, and which is largely derived from the detailed data presented above for TR upregulation during amphibian metamorphosis.

If upregulation of TR gene expression is a requisite for the consequential activation of target genes whose products characterize metamorphosis, then the simplest model would be one based on the dual threshold of receptor numbers. The existence of such a differential threshold of receptor concentration would explain both the receptor autoinduction and target gene activation by the hormone.[7,74,75] According to the simple model presented in Fig. 6.15, genes encoding receptors would be constitutively expressed to allow low levels of functional receptor to be present in all target tissues during early stages of development. The unliganded receptor would be inactive or, as in the case of TR, may even keep the downstream target genes in a repressed state. Upon the maturation of the thyroid gland and secretion of thyroid hormone, it is proposed that the low levels of activated receptor would suffice to upregulate the expression of receptor genes but not that of the target genes. For the latter process, and as the secretion of thyroid hormone increases (see chapter 5), concentrations of liganded TR above a threshold higher than that necessary for TR autoinduction would be required to activate different sets of genes in different tissues, each depending on distinct developmental programmes. Besides the fact that definitive experimental evidence is still awaited, an intriguing question arises as to the factors involved in maintaining low levels of receptor prior to the onset of, and high levels during, postembryonic development. One such factor could be the involvement of other hormones or signaling molecules. Hormonal cross-regulation of nuclear receptor is the topic of the chapter that follows.

References

1. Schulster D, Levitzki A. Cellular Receptors. Chichester, UK: John Wiley, 1980.
2. Eriksson H, Gustafsson J-A. Steroid Hormone Receptors: Structure and Function. Nobel Symposium No. 57. Amsterdam: Elsevier; 1983.
3. Baulieu E-E, Kelly PA. Hormones. From Molecules to Disease. Paris: Hermann, 1990.
4. Parker MG. Nuclear Hormone Receptors. London: Academic Press. 1991.
5. Kahn CR, White MF, Shoelson SE et al. The insulin receptor and its substrate: molecular determinants of early events in insulin action. Rec Progr Horm Res 1993; 48:291-339.
6. Kaplan J. Polypeptide-binding membrane receptors: Analysis and classification. Science 1981; 212:14-20.
7. Tata JR. Autoregulation and crossregulation of nuclear receptor genes. Trends Endocrin Metab 1994; 5:283-90.
8. Tata JR, Baker BS, Machuca I et al. Autoinduction of nuclear receptor genes and its significance. J Ster Biochem Mol Biol 1993; 46:105-19.
9. Tata JR. Hormonal interplay and thyroid hormone receptor expression during amphibian metamorphosis. In: Gilbert LI, Tata JR,

Atkinson BG, eds. Metamorphosis. Postembryonic Reprogramming of Gene Expression in Amphibian and Insect Cells. San Diego, Academic Press. 1996b:465-503.

10. Tata JR, Smith DF. Vitellogenesis: a versatile model for hormonal regulation of gene expression. Recent Prog Horm Res 1979; 35:47-95.

11. Wahli W, Dawid IB, Ryffel GU et al. Vitellogenesis and the vitellogenin gene family. Science 1981; 212:298-304.

12. Shapiro DJ, Barton MC, McKearin DM et al. Estrogen regulation of gene transcription and mRNA stability. Rec Progr Horm Res 1989; 45:29-58.

13. Schimke RT, Rhoads RE, Palacios R et al. Ovalbumin mRNA, complementary DNA and hormone regulation in chick oviduct. Karolinska Symposia on Research Methods in Reproductive Endocrinology 1973; 6:357-75.

14. O'Malley BW, Tsai SY, Bagchi M et al. Molecular mechanism of action of a steroid hormone receptor. Rec Progr Horm Res 1991; 47:1-24.

15. Westley B, Knowland J. An estrogen receptor from *Xenopus* laevis liver possibly connected with vitellogenin synthesis. Cell 1978; 15:367-75.

16. Wolffe AP, Tata JR. Coordinate and non-coordinate estrogen-induced expression of A and B groups of vitellogenin genes in male and female *Xenopus* hepatocytes in culture. Eur J Biochem 1983; 130:365-72.

17. Ng WC, Wolffe AP, Tata JR. Unequal activation by estrogen of individual *Xenopus* vitellogenin genes during development. Dev Biol 1984; 102:238-47.

18. Perlman AJ, Wolffe AP, Champion J et al. Regulation by estrogen receptor of vitellogenin gene transcription in *Xenopus* hepatocyte cultures. Mol Cell Endocrin 1984; 38:151-61.

19. Barton MC, Shapiro DJ. Transient administration of estradiol-17β establishes an autoregulatory loop permanently inducing estrogen receptor mRNA. Proc Natl Acad Sci USA 1988; 85:7119-23.

20. Pakdel F, Le Guellec C, Vaillant C et al. Identification and estrogen induction of two estrogen receptors (ER) messenger ribonucleic acids in the rainbow trout liver: Sequence homology with other ERs. Mol Endocrin 1989; 3:44-51.

21. Lerivray H, Smith JA, Tata JR. FOSP-1 (Frog oviduct specific protein-1) gene: Cloning of cDNA and induction by estrogen in primary cultures of *Xenopus* oviduct cells. Molec Cell Endocrin 1988; 59:241-48.

22. Varriale B, Tata JR. Autoinduction of estrogen receptor is associated with FOSP-1 mRNA induction by estrogen in primary cultures of *Xenopus* oviduct cells. Mol Cell Endocrin 1990; 71 R25-R31.

23. Tata JR. Hormonal and developmental regulation of *Xenopus* estrogen receptor and egg protein gene expression. In: Hochberg RB, Naftolin F. The New Biology of Steroid Hormones (Serono Symposium), New York: Raven Press, vol 74, 1991:213-25.

24. Tata JR. Early metamorphic competence of *Xenopus* larvae. Dev Biol 1968; 18:415-40.

25. Kawahara A, Baker B, Tata JR. Developmental and regional expression of thyroid hormone receptor genes during *Xenopus* metamorphosis. Development 1991; 112:933-43.

26. Leloup J, Buscaglia M. La triiodothyronine, hormone de la metamorphose des amphibiens. Comptes rendus hebdomadaires des Seances de l'Academie des Sciences, Paris 1977; 284:2261-63.

27. Shi Y-B, Wong J, Puzianowska-Kuznicka M, Stolow MA. Tadpole competence and tissue-specific temporal regulation of amphibian metamorphosis: roles of thyroid hormone and its receptor. BioEssays 1996; 18:391-99.

28. Yaoita Y, Brown DD. A correlation of thyroid hormone receptor gene expression with amphibian metamorphosis. Genes Dev 1990; 4:1917-24.

29. Rabelo EML, Baker B, Tata JR. Interplay between thyroid hormone and estrogen in modulating expression of their receptor and vitellogenin genes during *Xenopus* metamorphosis. Mech Develop 1994; 45:49-57.

30. Atkinson BG, Helbing C, Chen Y. Reprogramming of genes expressed in amphibian liver during metamorphosis. In: Gilbert LI, Tata JR, Atkinson BG, eds. Metamorphosis. Postembryonic Reprogramming of Gene Expression in Amphibian and Insect Cells. San Diego: Academic Press, 1996:539-66.

31. Gilbert LI, Tata JR, Atkinson, BG. Metamorphosis. Postembryonic Reprogramming of Gene Expression in Amphibian and Insect Cells. San Diego: Academic Press, 1996.

32. Machuca I, Tata JR. Autoinduction of thyroid hormone receptor during metamorphosis is reproduced in *Xenopus* XTC-2 cells. Molec Cell Endocr 1992; 87:105-13.

33. Kanamori A, Brown DD. The regulation of thyroid hormone receptor β genes by thyroid hormone in *Xenopus laevis*. J Biol Chem 1992; 267:739-45.

34. Kanamori A, Brown DD. Cultured cells as a model for amphibian metamorphosis. Proc Natl Acad Sci USA 1993; 90:6013-17.

35. Ulisse S, Esslemont G, Baker BS et al. Dominant-negative mutant thyroid hormone receptors prevent transcription from *Xenopus* thyroid hormone receptor beta gene promoter in response to thyroid hormone in *Xenopus* tadpoles in vivo. Proc Natl Acad Sci USA 1996; 93:1205-9.

36. Weber R. Biochemistry of amphibian metamorphosis. In: The Biochemistry of Animal Development. New York: Academic Press, 1967:227-301.

37. Cohen PP. Biochemical differentiation during amphibian metamorphosis. Science 1970; 168: 533-44.

38. Frieden E, Just JJ. Hormonal responses in amphibian metamorphosis. In: Litwack G, ed. Biochemical Actions of Hormones. New York: Academic Press, 1970:2-52.

39. Beckingham Smith K, Tata JR. The hormonal control of amphibian metamorphosis. In: Graham C, Wareing PF, eds. Developmental Biology of Plants and Animals. Oxford: Blackwell, 1976:232-45.

40. Safi R, Begue A, Hanni C et al. Thyroid hormone receptor genes of neotenic amphibians. J Mol Evoln 1997; 44:595-604.

41. Mangelsdorf DJ, Thummel C, Beato M et al. The nuclear receptor superfamily: the second decade. Cell 1995; 83:835-9.

42. Eliceiri BP, Brown DD. Quantitation of endogenous thyroid hormone receptors α and β during embryogenesis and metamorphosis in *Xenopus laevis*. J Biol Chem 1994; 269:24459-65.

43. Fairclough L, Tata JR. An immunocytochemical analysis of expression of thyroid hormone receptor α and β proteins during natural and thyroid hormone-induced metamorphosis in *Xenopus*. Dev Growth Differn 1997; 39:273-83.

44. Lechan RM, Qu Y, Berrodin TJ et al. Immunocytochemical delineation of thyroid hormone receptor β2-like immunoreactivity in the rat central nervous system. Endocrinology 1993; 132:2461-9.

45. Shi Y-B, Yaoita Y, Brown DD. Genomic organization and alternative promoter usage of the two thyroid hormone receptor beta genes in *Xenopus laevis*. J Biol Chem 1992; 267:733-8.

46. Ranjan M, Wong J, Shi Y-B. Transcriptional repression of *Xenopus* TRβ gene is mediated by a thyroid hormone response element located near the start site. J Biol Chem 1994; 269:24699-705.

47. Machuca I, Esslemont G, Fairclough L, Tata JR. Analysis of structure and expression of the *Xenopus* thyroid hormone receptor β (xTRβ) gene to explain its autoinduction. Mol Endocrin 1995; 9:96-108.

48. Baniahmad A, Tsai SY, O'Malley BW et al. Kindred thyroid hormone receptor is an active and constitutive silencer and a repressor for thyroid hormone and retinoic acid responses. Proc Natl Acad Sci USA 1992; 89:10633-7.

49. Barettino D, Bugge TH, Bartunek P et al. Unliganded T3R, but not its oncogenic variant v-erbA, suppresses RAR-dependent transactivation by titrating out RXR. EMBO J 1993; 12:1343-54.

50. Lazar MA. Thyroid hormone receptors: multiple forms, multiple possibilities. Endocr Rev 1993; 14:184-93.

51. Fondell JD, Roy AL, Roeder RG. Unliganded thyroid hormone receptor inhibits formation of a functional preinitiation complex: implications for active repression. Genes Dev 1993; 7:1400-10.

52. Forrest D, Golarai G, Connor J et al. Genetic analysis of thyroid hormone receptors in development and disease. Rec Progr Horm Res 1996; 52:1-22.

53. Chatterjee VKK, Nagaya T, Madison LD et al. Thyroid hormone resistance syndrome: Inhibition of normal receptor function by mutant thyroid hormone receptors. J Clin Invest 1991; 87:1977-84.

54. Refetoff S, Weiss RE, Usala SJ. The syndromes of resistance to thyroid hormone. Endocrin Rev 1993; 14:348-99.

55. Yen PM, Chin WW. New advances in understanding the molecular mechanisms of thyroid hormone action. Trends Endocrin Metab 1994; 5:65-72.

56. Zenke M, Munoz A, Sap J et al. V-erbA oncogene activation entails the loss of hormone-dependent regulatory activity of c-erbA. Cell 1990; 61:1035-49.

57. Brown TR. Androgen receptor dysfunction in human androgen insensitivity. Trends Endocrin 1995; 6:170-5.

58. Quigley CA, De Bellis A, Marschke KB et al. Androgen receptor defects: historical, clinical, and molecular perspectives. Endocrin Rev 1995; 16:271-321.

59. Minucci S, Zand DJ, Dey A et al. Dominant negative retinoid X receptor β inhibits retinoic acid-responsive gene regulation in embryonal carcinoma cells. Mol Cell Biol 1994; 14:360-72.

60. de Thé H, Vivanco-Ruiz M, Tiollais P et al. Identification of a retinoic acid responsive element in the retinoic acid receptor beta gene. Nature 1990; 343:177-80.

61. Kastner P, Mark M, Chambon P. Nonsteroid nuclear receptors: What are genetic studies telling us about their role in real life? Cell 1995; 83:859-69.

62. Lehmann JM, Zhang X-K, Pfahl M. RARγ2 expression is regulated through a retinoic acid response element embedded in Spl sites. Mol Cell Biol 1992; 12:2976-85.

63. Leid M, Kastner P, Chambon P. Multiplicity generates diversity in the retinoic acid signaling pathways. Trends Biochem Sci 1992; 17:427-433.

64. Chambon P. The molecular and genetic dissection of the retinoid signaling pathway. Rec Progr Horm Res 1995; 50:317-32.

65. Mora GR, Mahesh VB. Autoregulation of androgen receptor in rat ventral prostate: involvement of *c-fos* as a negative regulator. Mol Cell Endocrin 1996; 124:111-20.

66. Yen PM, Liu Y, Palvimo JJ et al. Mutant and wild-type androgen receptors exhibit cross-talk on androgen-, glucocorticoid-, and progesterone-mediated transcription. Mol Endocrin 1997; 11:162-71.

67. Lebel JM, L'Herault S, Dussault JH et al. Thyroid hormone upregulates thyroid hormone receptor beta gene expression in rat cerebral hemisphere astrocyte cultures. Glia 1993; 9:105-112.

68. Andres AJ, Thummel CS. Hormones, puffs and flies: the molecular control of metamorphosis by ecdysone. Trends Gen 1992; 8:132-8.

69. Thummel CS. From embryogenesis to metamorphosis: the regulation and function of Drosophila nuclear receptor superfamily members. Cell 1995; 83:871-77.

70. Thummel CS. Flies on steroids—*Drosophila* metamorphosis provides insights into the molecular mechanisms of steroid hormone action. Trends Gen 1996; 12:306-10.

71. Karim FD, Thummel CS. Temporal coordination of regulatory gene expression by the steroid hormone ecdysone. EMBO J 1992; 11:4083-93.

72. Ashburner M, Chihara C, Meltzer P et al. Temporal control of puffing activity in polytene chromosomes. Cold Spring Harb Symp Quant Biol 1974; 38:655-62.

73. Russell S, Ashburner M. Ecdysone-regulated chromosome puffing in *Drosophila melanogaster*. In: Gilbert LI, Tata JR, Atkinson BG eds. Metamorphosis. Postembryonic Reprogramming of Gene Expression in Amphibian and Insect Cells. San Diego: Academic Press, 1996:109-44.

74. Tata JR. Amphibian metamorphosis: An exquisite model for hormonal regulation of postembryonic development in vertebrates. Develop Growth Differ 1996a; 38:223-31.

75. Tata JR. Hormonal signaling and amphibian metamorphosis. Adv Dev Biol 1997; 5: 237-74.

76. Xu Q, Tata JR. Characterization and developmental expression of *Xenopus* C/EBP gene. Mech Dev 1992; 38:69-81.

CHAPTER 7

Hormonal Crossregulation of Nuclear Receptors

A s increasingly complex growth and developmental systems evolved, their regulation became correspondingly less and less simple. This complexity is characterized by the participation of multiple signals working through more than one signaling pathway in order to regulate a common physiological function. It is therefore not surprising that most developmental processes in higher eukaryotes are regulated by multiple hormones. The multiplicity may be manifested as a mutual potentiation or an inhibition of the action of any individual hormone or signaling molecule. The importance of auto-upregulation of a single nuclear receptor by a unique hormone has been considered in the preceding chapter. Here the same concept is now extended to an examination of hormonal crossregulation of nuclear receptors, involving both up- and downregulation.

Multihormonal Growth and Developmental Systems

When the role of the pituitary and hypothalamus as master glands orchestrating the function of other vertebrate endocrine tissues became clear, the issues of homeostasis and feedback mechanisms assumed increasing importance. The same was extended to the central control of the invertebrate endocrine system (for extensive reviews see refs. 1-3). For example, an overproduction of thyroid (TH) or glucocorticoid hormone would be rapidly adjusted by these hormones acting on the pituitary and suppressing the production of thyrotropin (TSH) or corticotropin (ACTH), respectively. This type of feedback can operate either by inhibiting the activities of the pituitary directly or via the hypothalamus, or both (see Fig. 7.1). Direct feedback inhibition by the end-hormones of these two CNS tissues has been repeatedly demonstrated in organ culture studies. This negative regulation is particularly important for adult organisms to

Hormonal Signaling and Postembryonic Development,
by Jamshed R. Tata. © 1998 Springer-Verlag and R.G. Landes Company.

prevent under- or over-secretion of hormones and, thus, that of hypo- or hyper-physiologic responses to a given hormone. For example, the resultant homeostasis is important for the maintenance of the basal metabolic rate in mammals, and that of mineral and water balance and regular ovulation in all vertebrates.

At the molecular level feedback inhibition operating between two endocrine organs is best explained by negative regulation at the level of cell and receptor specificities. At the cellular level different cell-types in the anterior pituitary respond separately to the different releasing hormones from the hypothalamus. Thus as indicated in Figure 7.1, TRH, CRF and GnRH exert their stimulatory action, via membrane receptors in thyrotropes, corticotropes and gonadotropes, on the synthesis and secretion of the tropic hormones TSH, ACTH and LH/FSH, respectively. The latter, also via membrane receptors, activate the thyroid, adrenals and gonads, respectively, to increase the production of T_4/T_3, GC and sex steroids. These hormones are recognized by the cognate nuclear receptors in the respective pituitary cell-types where they exert a negative or inhibitory action. Any over-production of these hormones would result in a diminution of the production of the hypophyseal tropic hormones (or even hypothalamic releasing factors).[3]

Recent studies on thyroid hormone receptors in the pituitary and hypothalamus have revealed the presence of negative response elements in the promoters of TSH and TRH genes with which the TRs interact to diminish the expression of the respective genes encoding their precursors. As already discussed in chapter 3 (see section 3.2; Fig. 3.10),[4] the TRE that regulates the expression of TSH gene is quite distinct from the positive TREs in genes that activated by T_3. At the same time, there is evidence for negative specificity at the level of the receptor. It has been suggested that a distinct isoform of TR (TRβ2) which is only present in the pituitary specifically interacts with the negative TRE(s) of the TSH gene upon binding its ligand.[4] Because TR functions as a heterodimer with RXR (see chapter 3) it is interesting to note that in a recent study[5] RXRs were found to enhance the T_3 suppression of transactivation of the human TSH gene on negative TREs in the promoter of this gene. Thus, negative feedback loops, which are of particular importance in clinical endocrinology are established by crossregulation of hormone receptors.

Another well-studied multihormonal regulatory system is that of the development of the mammary gland and the onset and main-

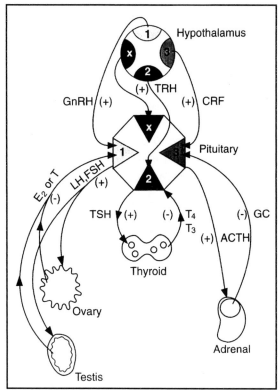

Fig. 7.1. Crossregulation via membrane and nuclear hormone receptors is the basis for feedback inhibition loops between the pituitary and its target endocrine organs. The hypothalamic releasing hormones GnRH, TRH and CRF act on respective membrane receptors in distinct populations of cells in the anterior pituitary, i.e., gonadotropes (1), thyrotropes (2), corticotropes (3), to stimulate the release of LH/FSH, TSH and ACTH, respectively. The latter, in turn, act on membrane receptors in the gonads, thyroid and adrenals to enhance the synthesis and secretion of gonadal steroids T or E_2, T_3/T_4 and GC, respectively. These terminal hormones can suppress the production of pituitary tropic hormones by acting on specific negative response elements in their genes by combining with the appropriate nuclear receptors in different pituitary cells 1, 2 and 3. This negative crossregulation completes the feedback loop and ensures the controlled production of hormones by peripheral endocrine glands. Abbreviations: GnRH, gonadotropin releasing hormone; TRH, thyrotropin (TSH) releasing hormone; CRF, corticotropin (ACTH) releasing factor; T, testosterone; E_2, estrogen; LH, luteotropin; FSH, follicle stimulating hormone; T_4/T_3, thyroid hormones; GC, glucocorticoid. x represents other hypothalamopituitary systems.

tenance of lactation. As already discussed in chapter 2, during pregnancy the functional development of the mammary gland is dependent on the interplay between insulin, estrogen (and progesterone) glucocorticoid and prolactin (see Fig. 2.7). Numerous studies in vivo and on organ and cell cultures in a variety of mammalian tissues have established that the potentiating action of these hormones requires strict sequential and concomittant interactions. The cloning of genes encoding specific milk proteins, such as caseins and lactalbumin, have enabled investigators to carry out molecular analysis of the facilitative and inductive roles of these hormones and their temporally regulated interactions during all stages of mammary gland development and function.[6] These studies have revealed that the regulation of expression of milk protein genes by prolactin is dependent on the precise sequential interplay among all the hormones involved. However, despite the fact that the membrane receptors for insulin and prolactin, and the nuclear receptors for estrogen and glucocorticoids including their transduction systems, have been well-characterized (see chapter 3). Relatively little is known about how this hormonal interplay is effected at the level of the receptor. What is certain is that this type of multihormonal dependency must involve crossregulation of receptor for one hormone by that of another.

Crossregulation of Nuclear Receptors

In the preceding chapter we introduced the topic of autoregulation of receptor expression with downregulation of progesterone receptor (PR) by progesterone in the uterus (Fig. 6.1). However, in the same experiments, estrogen, which promotes uterine growth, had the opposite effect in that it substantially upregulated the level of PR.[7] The discussion below of crossregulation of nuclear receptors during postembryonic development is largely based on work from the author's laboratory.[8–10]

Thyroid and Glucocorticoid Hormones Potentiate the Autoinduction of ER During Vitellogenesis

One of the first indications, albeit indirect, that thyroid hormone may modify the response of a tissue to estrogen was obtained from our studies on the activation of vitellogenin genes in *Xenopus* liver, designed to determine when during development would the organism first respond to E_2. As determined by de novo induction of Vit gene transcription, the onset of response was found to be associated with that of metamorphosis[11] (see Fig. 7.2). Since metamorphosis is initiated by the secretion of endogenous thyroid hormones (see chapter 5), one explanation for this ontogeny of responsiveness to exogenous estrogen would be that TH was somehow responsible for this early acquisition of ability to express Vit genes at the onset of metamorphosis. Quite significantly, administration of T_3 to pre-metamorphic *Xenopus* tadpoles, which induces metamorphosis precociously, also advanced the induction of Vit genes by E_2 at early tadpole stages, i.e., the curve in Fig. 7.2 was shifted to the left (data not shown). Other workers had also suggested that thyroid hormone had

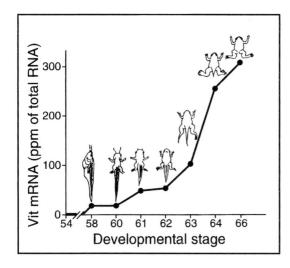

Fig. 7.2. Acquisition of competence to express vitellogenin (Vit) genes in response to exogenous estrogen (E_2) following the onset of metamorphosis. Batches of *Xenopus* tadpoles and froglets at the developmental stages indicated (54-66) were exposed to 1 μM E_2 in their water for 3 days before extraction of RNA from the liver. The concentration of four Vit mRNAs was measured as described by Ng et al.[11] Note that endogenous thyroid hormone is first secreted around stage 53/54, reaching a maximum at stages 60-62 (metamorphic climax) and drops to low levels by froglet stage 66 (see Fig. 6.4).

a role to play in rendering vitellogenin inducible by estrogen during *Xenopus* tadpole development and metamorphosis.[12-14] In view of the close association between the de novo activation of Vit genes and the upregulation of ER in primary cultures of male *Xenopus* hepatocytes[15] (see chapter 6 and Fig. 6.2), it was thought important to determine whether T_3 would affect the expression of both ER and Vit genes in these cultures.

Wangh and Schneider[16] had demonstrated that the addition of T_3 to adult male *Xenopus* hepatocyte cultures facilitated the inducibility of vitellogenin by E_2. The same was shown to be true for larval hepatocytes.[17] To determine whether T_3 exerted this action by also modifying the autoinduction of ER by E_2, Rabelo and Tata[18] measured the accumulation of ER transcripts in these cultures in the presence of different combinations of T_3 and E_2. As shown previously by RNase protection assays, E_2 upregulated the expression of *Xenopus* ER mRNA in male hepatocytes (see chapter 6, section 6.2). If these cells were also exposed to T_3 then the autoinduction of ER transcripts was substantially enhanced.[18] Later, similar experiments were repeated in whole *Xenopus* tadpoles at different stages.[19] As can be seen in Fig. 7.3, the crossinduction of ER was found to be developmental stage specific. In line with earlier studies,[20-22] T_3 was found to autoinduce TR mRNA in several tissues of stage 52 *Xenopus* tadpoles. However, it did not enhance the upregulation by E_2 of ER in the livers of stage 56 tadpoles (onset of natural metamorphosis), but only

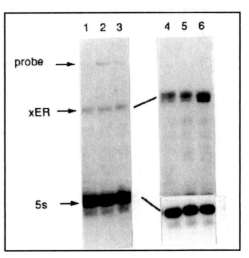

Fig. 7.3. T_3 enhances the auto-induction of ER mRNA by E_2 in *Xenopus* tadpole liver in a developmentally regulated manner. ER mRNA (xER) was determined by RNase protection assay in livers from stage 56 (lanes 1-3) and 60 (lanes 4-6) tadpoles. 5S RNA was used as an internal standard. Lanes 1, 4: control (untreated) tadpoles; 2, 5: treated for 5 days with 0.1 μM E_2; 3, 6: tadpoles pre-treated with 1 nM T_3 for 1 day followed by T_3 and E_2 for the next 4 days. Other details in ref. 19.

did so in tadpoles at metamorphic climax (stage 60). This result suggests that some developmentally regulated factors other than those that are T_3-inducible may also play a role in establishing estrogen sensitivity during the progression of metamorphosis.

Would the enhancement by T_3 of autoinduction of ER mRNA lead to more ER protein and, consequently, to higher level of Vit gene activation by E_2? The next step was therefore to determine the activation of Vit genes both in cell cultures and in vivo. Such experiments indeed revealed that the crossregulation of ER expression by T_3 underlined a simultaneous increase in the accumulation of ER and Vit mRNAs. A good visual demonstration of this finding was provided by the in situ hybridization of Vit mRNA in the liver of intact *Xenopus* tadpoles which had nearly completed metamorphosis (stage 62/63) and were hormonally manipulated. Fig. 7.4 shows that in the controls the vitellogenin genes are silent in this tissue in late metamorphosis. Exposure of tadpoles to 10 μM E_2 alone for 1 day fails to elicit a signal for Vit mRNA but at 2 days a low but significant signal can be discerned (Fig. 7.4c and e). Pre-treatment with T_3 for 1 or 2 days produced the same intense Vit mRNA signal as after 4 days of treatment with E_2 alone (Figs. 7.4f, g, h). Not shown in this figure is that T_3 alone did not induce Vit mRNA at any time. Thus, thyroid hormone not only potentiates the induction of Vit mRNA by estrogen but it also promotes an earlier activation of the Vit genes by the steroid hormone. What is particularly important is that, under the same experimental conditions, T_3 fails to influence the autoinduction of ER at the onset of metamorphosis, whereas it

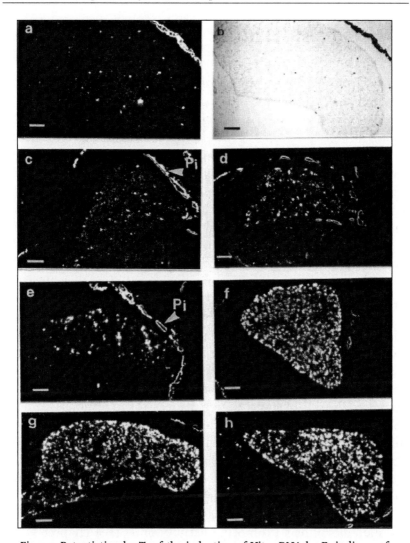

Fig. 7.4. Potentiation by T_3 of the induction of Vit mRNA by E_2 in livers of *Xenopus* tadpoles at late stages of metamorphosis. Vit mRNA is visualized by in situ hybridization in saggital sections of stage 63/64 tadpoles treated with 0.1 µM E_2 alone or after pre-treatment with 1 nM T_3 for different periods of time. Only the dark-field images with antisense Vit cRNA are shown as the sense probes gave no signal. a: Control liver; b: bright-field image of control; c: 1 day treatment with E_2 alone; d: 1 day pre-treatment with T_3 followed by E_2 for 1 day; e: 2 days E_2 alone; f: 1 day T_3 followed by 2 days E_2; g: 4 days E_2 alone; h: 2 days T_3 followed by 2 days E_2. Where T_3 was used, it was present in the water throughout the period of exposure to E_2. Arrows indicate pigmentation layer surrounding the liver. Bars indicate 100 µM. Other details in ref. 19.

strongly potentiates the response to E_2 at metamorphic climax (Fig. 7.3). Thus, a developmentally regulated interplay between thyroid hormone and estrogen, mediated by auto- and cross-regulation of their receptor genes, determines the kinetics and magnitude of activation of vitellogenin genes during postembryonic development.

Besides thyroid hormone, glucocorticoids also potentiate the expression of egg protein genes in the oviduct and liver of oviparous vertebrates[23-26] (see chapter 4). Wangh[27] had described that glucocorticoids act together with thyroid hormone in regulating estrogen-induced vitellogenesis in *Xenopus* liver. It was therefore of some interest to examine the interplay between these three hormones as a function of auto- and cross-regulation of their receptors. In experiments undertaken to explore this issue, Ulisse and Tata[28] observed that T_3 and dexamethasone (Dex) both rapidly induced ER mRNA by independent pathways in primary cultures of adult male *Xenopus* hepatocytes (data not shown). This crossregulation of ER mRNA is accompanied by a marked enhancement of the activation of the silent vitellogenin B1 gene if estradiol is added 12 hours after T_3 and Dex, thus indicating an elevated level of functional ER induced by the two hormones. This conclusion was supported by a separate experiment which showed that pre-incubation with the two hormones increased the rate of transcription of an estrogen response element-CAT (ERE-CAT) construct transfected into cultured hepatocytes before addition of estrogen. T_3 and Dex exert additive effects, with different kinetics, on the induction of ER and Vit mRNAs by E_2, a finding which emphasizes the importance of nuclear receptor expression in the interplay in multiple hormonal signaling.

Exogenous prolactin (PRL) exerts a strong inhibitory effect on the metamorphosis-promoting activity of T_3 (see chapter 5). Although no information was available for some time as to whether adult amphibian liver would respond to PRL, it was of some interest to determine if PRL would modify the potentiation by T_3 of autoinduction of ER and the consequent activation of Vit genes. As shown in Fig. 7.5, PRL did indeed exert a strong effect on the action of T_3 in primary cultures of adult male *Xenopus* hepatocytes. Briefly, PRL on its own does not activate the Vit genes nor modify the steady-state levels of ER and TR mRNAs. As already noted earlier (Fig. 7.3), T_3 alone did not induce ER or activate Vit genes. Significantly, PRL blocked the potentiation by T_3 of the auto-induction of ER mRNA and activation of Vit genes by E_2, in addition to the upregulation of

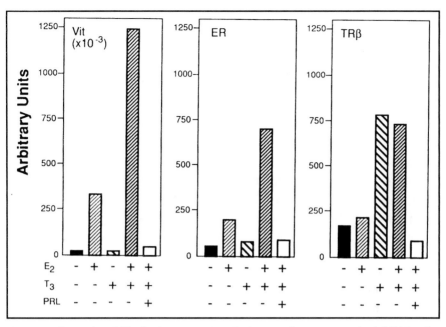

Fig. 7.5. Prolactin (PRL) blocks the expression of Vit, ER and TRβ mRNAs by inhibiting the potentiation by T_3 of the autoinduction of ER mRNA and activation of vitellogenin (Vit) genes by estrogen (E_2). The histograms show relative amounts of steady-state levels of Vit, ER and TRβ mRNAs in primary cultures of male *Xenopus* hepatocytes treated with different combinations of 0.1 μM E_2, 2 nM T_3 and 0.5 i.u./ml of PRL for 24 hours before the extraction of RNA and determination of concentrations of Vit, ER and TRβ mRNAs. The determinations of different mRNAs were made in the same RNA samples, and the relative units refer to each transcript separately. Other details as in refs. 18 and 19.

TR. This result emphasizes how multiple signals can be integrated into a network of nuclear receptor-mediated responses. Adult mammalian liver is well-known to express prolactin receptor (see chapter 3).[3,29] The clear negative effect of PRL on the activation of Vit genes is the first demonstration that adult amphibian liver responds to prolactin and, hence, expresses its receptor.

Whereas the function of thyroid hormone in regulating metabolic activity in adult mammals and homeotherms is one of the fundamentals of endocrinology (see chapter 2), what role this hormone plays in adult amphibian physiology has been debatable. The potentiation by T_3 of vitellogenin gene activation in adult *Xenopus* liver cells[18,19,27,28] raises the possibility that thyroid hormone also serves a physiological function in adult amphibia. That this involves a reproductive function has now to be considered. In this context, it is significant that circulating T_3 can be detected in adult *Xenopus*, albeit at a lower concentration than during metamorphosis.[8]

Crossregulation of TR Expression by Prolactin During Metamorphosis

All thyroid hormone-induced developmental responses in amphibia are inhibited or greatly diminished by exogenous prolactin or prolactin-like growth factors in vivo and in organ cultures (see chapter 5, Fig. 5.4).[8,30-33] The "juvenilizing" action of PRL allows one to "freeze" experimentally the process of natural or TH-induced metamorphosis by simply adding it to the tadpole's water or to the culture medium. This antimetamorphic action of PRL has thus been exploited in the author's laboratory to explore further the autoinduction of TRs and its significance.[8,19,34,35] For example, we have already seen in Fig. 5.4 the inhibition by PRL of both the actions of T_3 promoting the morphogenesis of limb buds and regression of *Xenopus* tadpole tails in culture in a manner similar to that in vivo.

When the autoinduction of TR mRNAs was studied in vivo and in vitro, PRL almost totally abolished the upregulation of transcripts of both isoforms by T_3. This inhibitory effect can be visualized for TRα mRNA in Fig. 7.6. In the presence of PRL the strength of the T_3-upregulated TRα mRNA signal was considerably reduced in all tissues.[8] It is particularly noticeable for brain, liver and intestine in the saggital sections in Fig. 7.6 of stage 54 *Xenopus* tadpoles that were hormonally manipulated. The same action can be reproduced at early stages of pre-metamorphic tadpoles (stage 50) when biochemical analysis revealed the loss of autoinduction of both TRα and β mRNAs (data not shown).[34] PRL alone does not affect the low, basal levels of TR mRNAs, but prevents their upregulation by T_3. This inhibition of TR autoinduction is quantitatively depicted in Table 7.1. The fact that the effect of PRL was manifested in early pre-metamorphic tadpoles suggests that PRL receptors are also expressed (as are TRs) well before natural metamorphosis commences.

That autoinduction of TR mRNAs leads to enhanced expression of TR proteins, which then provoke the action of T_3, was supported by the findings that PRL prevented the activation by T_3 of two well-documented target genes of amphibian metamorphosis listed in Table 5.2, namely serum albumin and 63 kDa adult-type keratin. This result is illustrated for *Xenopus* 63 kDa keratin mRNA

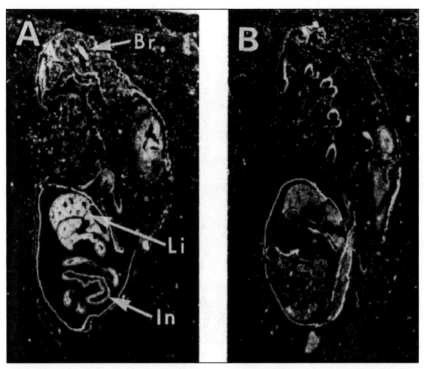

Fig. 7.6. Prolactin prevents the auto-upregulation of TRα mRNA, visualized by in situ hybridization. Stage 54 *Xenopus* tadpoles were treated with A) 1 nM T_3 alone for 3 days or B) with 0.1 i.u./ml of PRL for 4 days. Saggital sections of tadpoles (after removal of the tail) were hybridized to ^{35}S-labelled antisense TRα cRNA and the mRNA visualized by autoradiography. Br, brain; Li, liver; In, intestinal epithelium. From Rabelo EML, Baker B, Tata JR. Interplay between thyroid hormone and estrogen in modulating expression of their receptor and vitellogenin genes during *Xenopus* metamorphosis. Mech Develop 1994; 45:49-57.

in Fig. 7.7. The obligatory switching from the larval to the adult-type 63 kDa keratin gene by T_3 during metamorphosis has been well documented by Miller's laboratory.[36,37] If these experiments were prolonged then administration of PRL was also seen to block all the morphological aspects of metamorphosis. Thus, although it could be argued that the action of PRL in metamorphosis is indirect, the different pieces of evidence put together strongly suggest that TR autoinduction is a requirement for the induction and maintenance of postembryonic development by thyroid hormone.

How does prolactin inhibit the action of thyroid hormone? To answer that question, as a first step it is essential to have information about *Xenopus* prolactin receptor (PRLR) and the signal transduction pathway to the nucleus before one can explain its role in

Table 7.1. Prolactin (PRL) blocks the accumulation of TRα and β mRNAs in early stages of Xenopus tadpoles treated with T₃

Treatment	Relative units	
	TRα	TRβ
None	505	24
T3	1290	368
T3 + PRL	799	<10
PRL	405	43

Stage 50 tadpoles were exposed to 2 nM T₃ and/or 0.1 i.u. PRL/ml as indicated, for 4 days, before total RNA was extracted from whole larvae and the relative amounts of TRα and β estimated by quantitative RNase protection assays.

metamorphosis. But if it is assumed, as is most likely to be the case, that the receptor is a highly conserved evolutionarily, then the recent findings on mammalian PRLR and signal transduction pathway are some significance. As already mentioned in chapter 3 (Fig. 3.4), mammalian PRLR has been cloned and characterized as a relatively simple single transmembrane spanning receptor similar to cytokine receptors.[29,38,39] In a manner similar to cytokine receptors, liganded PRLR is thought to initiate a cascade of reactions, including tyrosine phosphorylation, to modify transcription factor activity in the nucleus via the *Raf*MEK or MAPKK-MAPK pathway (Fig. 3.4) along with the *fyn*JAK2 pathway to activate STAT 91.[40-43]

How precisely the chain of events initiated by PRL in an amphibian tadpole cell block the action of TH still remains to be explained. A recent publication however offers an intriguing clue. Stöcklin et al[44] have described how PRL-activated Stat 5, which is a member of a family of signal transducers and activators of transcription, forms a complex with the nuclear glucocorticoid receptor (GR). The complex binds to DNA independently of the glucocorticoid response element (GRE) in the target gene and diminishes the response to glucocorticoid in a GRE-containing promoter. If this mechanism of attenuation by PRL of a nuclear receptor-mediated pathway applies to the interplay between PRL and TH in amphibian metamorphosis, then the model proposed in Fig. 7.8 is worth considering as a possible explanation. Although oversimplified, the model can be tested experimentally. It also emphasizes the requirement of upregulation of TR to activate the ultimate target genes of T₃ via a dual threshold of TR concentration—a lower level for its

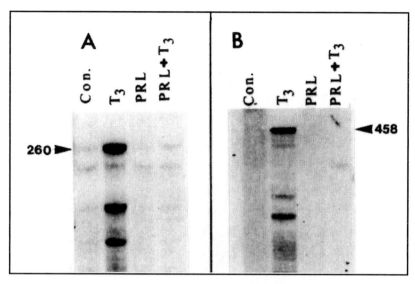

Fig. 7.7. Simultaneous inhibition by prolactin (PRL) of A) the autoinduction of TRβ mRNA and B) the activation de novo of adult-type 63 kDa keratin in pre-metamorphic *Xenopus* tadpoles treated for 4 days with 1 nM T_3 and 0.2 i.u. PRL/ml separately or together. RNA was extracted from stage 52 larvae and the two transcripts assayed in the same sample by RNase protection. The arrows mark the migration of the protected bands of 260 and 458 nucleotides for the TRβ and keratin transcripts, respectively. Con, control—no treatment. Other details in ref. 34.

autoinduction and higher level for the induction of 'adult' genes (see also Fig. 6.15). Different sets of downstream genes identified by different postembryonic developmental programs would then be regulated by a balanced interplay between positive and negative signals.

Interplay Between TH, GC and PRL and Co-Expression of TR and RXR

Nuclear receptors of the subgroup to which TR belongs, i.e., VDR, PPAR, RAR, etc., function in vivo as heterodimers formed with one or more isoforms of RXR (see chapter 3). It was therefore important to verify if and how the respective ligands of TR and RXR (T_3 and 9-cis-retinoic acid [9-cis-RA]) auto- and cross-regulate the expression of these two receptors. Since PRL and glucocorticoids exert opposing effects on the induction of metamorphosis by T_3 (see chapter 5) it would also be appropriate to examine how these two hormones influence TR autoinduction during metamorphosis. The studies described below serve to illustrate additional features of hormonal interplay during postembryonic development.

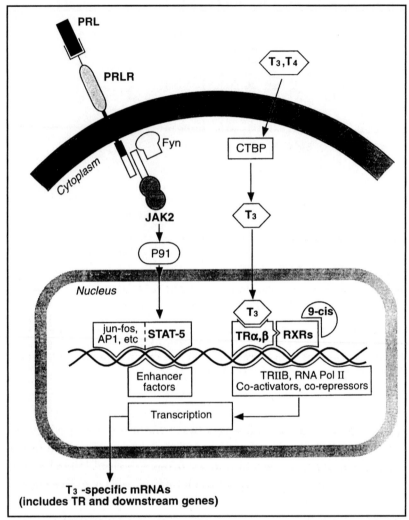

Fig. 7.8. A simplified mechanism to account for the crossregulation by prolactin (PRL) of the activation by TH (T_4, T_3) of its own receptor and downstream genes that characterize metamorphosis. According to this scheme the liganded PRL receptor (PRLR) in the cell membrane would activate the JAK/STAT signal transduction pathway leading to modification of the balance between STAT-like molecules (AP-1, Jun-Fos, etc.) and other transcription factors in the nucleus (see also chapter 3, and Fig. 3.4). T_4 and T_3 are thought to diffuse through the cell membrane and bind to cytoplasmic thyroid hormone binding proteins (CTBP), which determines the entry of TH into the cell nucleus.[68,69] T_3 binds to the low level of TRs and relieves the inhibition of transcription by the unliganded TRs α and β (see chapter 4). Among the first T_3 responsive genes are TRα and β, whose autoinduction would raise the concentration of these nuclear receptors to a higher threshold level to activate downstream genes (see chapter 6, Fig. 6.15). The activity of TRs is dependent on interactions with other nuclear receptors and transcription factors, such as heterodimerization with retinoid X receptors (RXRs) whose ligand is 9-cis-retinoic acid (9-cis), co-activators and co-repressors (see chapter 4). Among the latter would be STAT-like molecules generated by PRL which would diminish the positive action of liganded TRs.

Table 7.2. Quantitative representation of the data in Fig. 7.9 of the effect of hormonal treatment on the distribution of TRβ and RXRγ transcripts in cultured Xenopus *tadpole tails*

Hormonal Treatment	Relative mRNA concentration	
	TRβ	RXRγ
Control	1.0	1.0
T3	6.9	0.3
Dex	1.9	2.7
PRL	0.9	1.0
T_3 + Dex	11.5	0.7
T_3 + PRL	4.3	1.0
T_3 + Dex + PRL	5.6	1.1

The above values were obtained by densitometric scanning of the autoradiograms shown in Fig. 7.9, based on the concentrations of EF-1α RNA used as an internal control in each sample. Other details as in Fig. 7.9 and ref. 35.

In the preceding chapter we have seen *Xenopus* RXRα and γ form stable heterodimers with TRα and β, and which strongly bind DR+4 TREs of *Xenopus* TRβ gene promoter under conditions in which T_3 stimulates transactivation.[45,46] The first experiments therefore addressed the question of how the expression of both RXR and TR mRNAs in the same tadpole tissue was modulated by T_3 and other hormones. Since it is easier to manipulate tadpole tissues in culture than in whole larvae, organ cultures of *Xenopus* tadpole tails were treated with hormones for this purpose. It was thus found that prolactin inhibited, and the synthetic glucocorticoid dexamethasone potentiated, the T_3-induced regression of tadpole tail cultures[35] (see chapter 8). When the steady-state levels of TRα and β and RXRα and γ were determined in the same tissue sample, T_3 clearly upregulated TRβ mRNA in tail cultures, as it did in all other tissues (see chapter 6). But as shown in Fig. 7.9 and Table 7.2, quite unexpectedly T_3 strongly downregulated RXRγ transcripts (the same results were obtained for the α and β isomers of RXR). These opposing responses of mRNAs of the two heterodimeric receptor partners was in contrast to the effect of Dex. The glucocorticoid alone significantly increased the concentration of RXRγ mRNA, but when administered together with T_3 further enhanced the upregulation of TRβ, while at the same time diminishing the downregulation of RXR transcripts by T_3. Prolactin alone did not alter the steady-state levels of either receptor

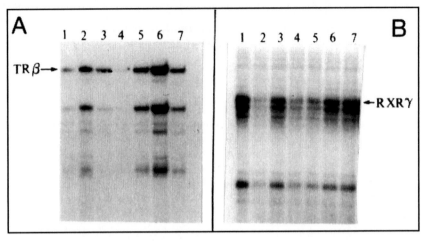

Fig. 7.9. Contrasting pattern of expression of TR and RXR transcripts in stage 54 *Xenopus* tadpole tail cultures treated with T₃, Dex and PRL. Organ cultures of stage 54 *Xenopus* tadpole tails were maintained for 4 days with different combinations of hormones after which RNA was extracted and the concentrations of A) TRβ and B) RXRγ transcripts were determined by RNase protection assays in the same samples of RNA. Hormonal treatment: Lane 1: control (no treatment); 2: 2.5 nM T₃; 3: 0.1 μM Dex; 4: 0.25 i.u. PRL/ml; 5: T₃ + PRL; 6: T₃ + Dex; 7: T₃ + Dex + PRL. Other details are given in ref. 35.

RNA, but it diminished the extent of autoinduction of TRβ mRNA by T₃, both in the presence and absence of Dex. The effects of all three hormones were dose-dependent. The contrasting pattern of expression of TR and RXR, the two heterodimeric partners of functional receptor offer a novel explanation of why glucocorticoids potentiate the action of TH in amphibian metamorphosis.

When the interplay between T₃, Dex and PRL was studied as a function of time, each hormone exerted its optimum effect with different kinetics (Fig. 7.10), which suggested possible separate pathways for different hormones, a result also obtained with whole tadpoles. Different results were obtained in another study[47] in that T₃ upregulated both TR and RXR transcripts in *Xenopus* tadpole limb and tail. The reason for the discrepancy is not known, which may possibly be due to different isoforms of RXR transcripts being analyzed. If that is indeed the reason, then one has to reconsider the possibility of differential tissue-specific receptor isoform expression. To better understand the contrasting patterns of responses to the three hormones, these studies in tail cultures or whole tadpoles were extended to *Xenopus* XTC-2 cells.[35] In line with earlier studies,[48,49]

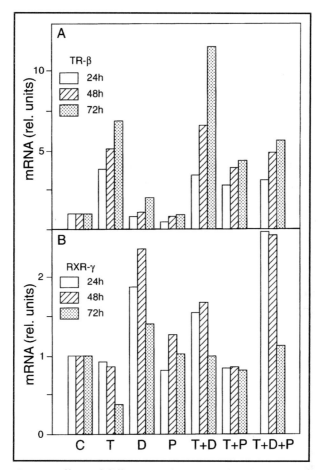

Fig. 7.10. Effects of different combinations of T_3 (T), Dex (D) and PRL (P), as a function of time, in organ cultures of *Xenopus* tadpole tails. Batches of tadpole tails were untreated (C) or treated with different hormones, as shown, for 24, 48 and 72 hours before the RNA was extracted and analyzed for the relative concentrations of TRα (A) and RXRβ (B) transcripts. Other details as in Fig. 7.9 and ref. 35.

the addition of 25 nM T_3 resulted in 5- and 50-fold increase in the accumulation of TRβ mRNA in these cells at 24 and 48 hours, respectively. In data not shown here, treatment with 0.1 μM Dex together with T_3 for 24 hours caused the steady-state levels to rise even further, i.e., from 20- to 150-fold. However, unlike in whole *Xenopus* tadpoles and organ cultures, PRL had very little effect on XTC-2 cells. These cells also contain steady-state levels of RXR transcripts that are too low to be meaningful for these studies, thereby not allowing a more precise analysis of the way in which the expression of the two nuclear receptor partners is regulated.

In later studies, Ulisse et al[50] compared the effects of T_3 and 9-cis-RA, the natural ligands of the two heterodimeric receptors, in XTC-2 cells. 9-cis-RA, but not T_3, produced a moderate upregulation of RXRα mRNA. As mentioned earlier, overexpression of TRβ in XTC-2 cells repressed basal transcription from a variety of TRE-containing promoters, which was relieved by the addition of T_3 (chapter 6).[45,46,51] Overexpression of RXRα and γ, on the other hand, failed to modify the expression of either TR or RXR. This result was unaltered irrespective of whether or not 9-cis-RA was added to the cells. It is not clearly established that exogenous 9-cis-RA gets into cells; however, co-transfection of XTC-2 cells with a RXRE-containing promoter and RXRα constructs revealed that these cells were capable of responding to exogenous 9-cis-RA. Thus, although XTC-2 cells contain functional RXR, 9-cis-RA does not seem to be required for TH-mediated transactivation. This finding raises the possibility worth considering in future studies that in these cultured amphibian cells TR may function as a homodimer.

Models Summarizing the Hormonal Interplay and Crossregulation of Nuclear Receptors

By putting together all the data obtained from various in vivo and in vitro manipulations, the two models depicted in Fig. 7.11 are proposed to summarize the interplay between some of the ligands described above and the cross-regulation of their respective receptors. The first model (Fig. 7.11.A) attempts to explain the interplay between E_2, T_3, GC and PRL leading to the activation of silent Vit genes in male *Xenopus* hepatocytes. A low level of ER is constitutively expressed in the naive liver cells which is considerably elevated by autoinduction by E_2 to reach a higher threshold sufficient to activate Vit genes. T_3 and GC both potentiate the upregulation of ER by its own ligand. They are thought to act by independent routes.[18,19,28] The action of T_3 is slow and is initiated by the upregulation of its own receptor, while that of GC is more rapid and may not require the upregulation of its own receptor, presumably because of a sufficiently high level of constitutively expressed GR in the liver.

In the second model of receptor crossregulation (Fig. 7.11.B), the activation of amphibian metamorphosis-specific genes is considered in terms of an interplay between T_3, PRL and GC. Metamor-

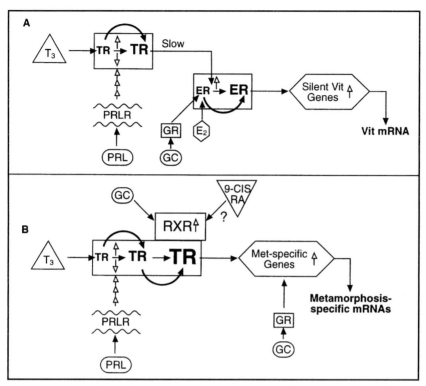

Fig. 7.11. Two models summarizing the crossregulation of expression of nuclear receptors by various hormones acting through both membrane and nuclear receptors, as exemplified by A) the activation of vitellogenin genes and B) metamorphosis-specific genes. In A) T_3 and GC potentiate the action of E_2 in activating the silent Vit genes in male *Xenopus* liver by upregulating the autoinduction of ER. PRL inhibits the potentiation by T_3 by blocking the autoinduction of TRs. In B) GC potentiates the action of T_3 in inducing metamorphosis by upregulating the expression of RXR but not of TR. PRL blocks the action of T_3 but inhibits autoinduction of TRs. The role of 9-cis-retinoic acid (9-cis-RA) is not clear. The straight upward and downward pointing arrows denote up and downregulation, respectively, while the curved, bold arrows represent autoinduction of nuclear receptors. Other explanations and abbreviations are in the text.

phosis is obligatorily dependent on T_3 which causes a massive upregulation of its own receptor, particularly TRβ, from its low constitutively expressed level from very early developmental stages (see above).[34,35,46] PRL prevents the autoinduction of TR, thereby diminishing or abolishing the induction of metamorphosis-specific genes. This action is a good example of cross-talk between signals transduced via membrane and nuclear receptors.[44,52-55] GC, on the other hand, potentiates the action of T_3 but does not cross-regulate the expression of TR. The glucocorticoid raises the level of transcripts of RXR which is required for heterodimerization with TR in order to activate the TRE-containing promoters of genes that specify the process of metamorphosis. Besides the interaction between TR and RXR being necessary for transactivation, there is now growing evidence for co-activators and co-repressors that modulate the transcriptional function of TR.[56-58] The number of such interactions between co-factors, some dependent on membrane-receptor-initiated phosphorylation cascade, and nuclear receptors will increase with time as details of the transcriptional machinery are further unravelled. Thus, in both examples depicted in Fig. 7.11, although one hormone is the key signal for a postembryonic developmental process, other hormones or signals can modulate the activation process by cross-regulation of the expression of nuclear receptors.

Crossregulation of Nuclear Receptors Is Widespread

Although the above examples have been selected on the basis of the interests of the author's laboratory with a view to analyzing the process in detail, crossregulation of other nuclear receptor expression is also well known.[3,9,59] Table 7.3 lists some examples of both up- and downregulation of nuclear receptors by hormones and other signals impinging on the cell membrane or nucleus. The upregulation of expression of ER by TH and GC, and the downregulation of TRα and β by PRL have already been described above. One of the best documented examples of receptor crossregulation is that of the upregulation by estrogen of progesterone receptor (PR) in various progesterone target tissues such as the immature chick oviduct and mammalian uterus and breast cancer cells.[3,55,60] This cross-talk explains the enhanced sensitivity to progesterone in the presence of estrogen. Two intriguing accounts of cross-talk between nuclear receptors reported recently and worth consideration are: 1) the interaction between androgen, glucocorticoid and progesterone receptors in activating the MMTV and tyrosine aminotransferase genes[61] and 2) the potentiation by TR of the PPAR/RXR-dependent activation of transcription of acyl-CoA oxidase gene in mouse liver.[62] An unusual example of multi-receptor cross-talk is the recent report that RXR and PPAR can activate estrogen responsive genes by their ability to bind directly to estrogen response element without the intervention of ER.[63]

Table 7.3. Crossregulation of expression and function of nuclear receptors

Hormone or intracellular signal	Receptor	Up or down regulation	Species and tissue	Function involved
TH	ER	Up	Xenopus liver	Vitellogenesis
E$_2$	PR	Up	Chick oviduct; human breast cancer	Egg maturation; cancer
P	ER	Up	Chick oviduct	Egg development
GC	ER	Up	Chick and frog liver	Vitellogenesis
PRL	TR	Down	*Xenopus* tadpole tissue	Metamorphosis
Insulin	PR	Up	Human breast cancer cells	Cancer
Cyclic AMP	ER	Up	Human breast cancer cells	Cancer
T,c-*Fos*	AR	Down	Rat prostate	Growth

P, progesterone; T, Testosterone or androgen; PR, progesterone receptor; AR, androgen receptor. Other abbreviations are as used routinely in this chapter.

Many instances of multihormonal regulation of nuclear receptor expression or function also involve cross-talk between membrane and nuclear receptors. The downregulation of TR by PRL has been discussed in detail above. Besides its upregulation by E$_2$, progesterone receptor is also regulated by insulin and insulin-like growth factor-I (IGF-I), both of which exert their action through membrane receptors and signaling via phosphokinases.[53,55] Importantly, the significance of the multifactor regulation of PR is that both E$_2$ and IGF-I are involved in the growth of human breast cancer cells. Among other recent examples of crossregulation via intracellular signals originating from the cell membrane are synergy between cyclic AMP and ER in human breast cancer cells[64] and the downregulation of androgen receptor in rat ventral prostate by c-*fos*.[65] Thus, it is clear that cross-talk among nuclear receptors and between membrane and nuclear receptors is quite widespread and not restricted to amphibian vitellogenesis and metamorphosis. The integration of signals via receptors is of great importance in both embryonic and postembryonic development. Since these receptors and many of the important intracellular intermediates are oncogene products or oncogene-derived products (see chapter 3),[66,67] the integration of signaling mechanisms emphasizes the central role of oncogenes in the evolution of signaling networks in developmental processes.

References

1. Turner CD, Bagnara J. General Endocrinology. Philadelphia: Saunders, 1971.
2. Barrington EJW. Chemical communication. Proc R Soc (London) Ser B 1977; 199:257.
3. Baulieu E-E, Kelly PA, eds. Hormones. From Molecules to Disease. Paris: Hermann, 1990.
4. Chatterjee VKK, Tata JR. Thyroid hormone receptors and their role in development. Cancer Surveys 1992; 14:147-67.
5. Takeda T, Nagasawa T, Miyamoto T et al. The function of retinoid X receptors on negative thyroid hormone response elements. Mol Cell Endocrinol 1997; 128:85-96.
6. Rosen JM, Matusik RJ, Richards DA et al. Multihormonal regulation of casein gene expression at the transcriptional and post-transcriptional levels. Rec Prog Horm Res 1980; 36:157-87.
7. Milgrom E. Steroid hormones. In: Baulieu E-E, Kelly PR, eds. Hormones. From Molecules to Disease. Paris: Hermann, 1990:386-437.
8. Tata JR, Baker BS, Machuca I et al. Autoinduction of nuclear receptor genes and its significance. J Ster Biochem Molec Biol 1993; 46:105-19.
9. Tata JR. Autoregulation and crossregulation of nuclear receptor genes. Trends Endocrin 1994; 5:283-90.
10. Tata JR. Hormonal interplay and thyroid hormone receptor expression during amphibian metamorphosis. In: Gilbert LI, Tata JR, Atkinson BG, eds. Metamorphosis. Postembryonic Reprogramming of Gene Expression in Amphibian and Insect Cells. San Diego: Academic Press, 1996:465-503.
11. Ng WC, Wolffe AP, Tata JR. Unequal activation by estrogen of individual *Xenopus* vitellogenin genes during development. Dev Biol 1984; 102:238-47.
12. Huber S, Ryffel GU, Weber R. Thyroid hormone induces competence for oestrogen dependent vitellogenin synthesis in developing *Xenopus* liver. Nature 1979; 278:65-7.
13. May FEB, Knowland J. Oestrogen receptor levels and vitellogenin synthesis during development of *Xenopus laevis*. Nature 1981; 292:853-5.
14. Knowland J. Induction of vitellogenin synthesis in *Xenopus laevis* tadpoles. Differentiation 1985; 12:47-51.
15. Perlman AJ, Wolffe AP, Champion J et al. Regulation by estrogen receptor of vitellogenin gene transcription in *Xenopus* hepatocyte cultures. Mol Cell Endocrin 1984; 38:151-61.
16. Wangh LJ, Schneider W. Thyroid hormones are corequisites for estradiol-17 in vitro induction of *Xenopus* vitellogenin synthesis and secretion. Dev Biol 1982; 89:287-93.
17. Kawahara A, Kohara S, Amano M. Thyroid hormone directly induces hepatocyte competence for estrogen-dependent vitellogenin synthesis during the metamorphosis of *Xenopus laevis*. Dev Biol 1989; 132:72-80.

18. Rabelo EM, Tata JR. Thyroid hormone potentiates estrogen activation of vitelloenin genes and autoinduction of estrogen receptor in adult *Xenopus* hepatocytes. Cell Endocrinol 1993; 96:37-44.
19. Rabelo EML, Baker B, Tata JR. Interplay between thyroid hormone and estrogen in modulating expression of their receptor and vitellogenin genes during *Xenopus* metamorphosis. Mech Develop 1994; 45:49-57.
20. Yaoita Y, Brown DD. A correlation of thyroid hormone receptor gene expression with amphibian metamorphosis. Genes & Dev 1990; 4:1917-24.
21. Kawahara A, Baker B, Tata JR. Developmental and regional expression of thyroid hormone receptor genes during *Xenopus* metamorphosis. Development 1991; 112:933-43.
22. Baker BS, Tata JR. Accumulation of proto-oncogene c-*erb*-A related transcripts during *Xenopus* development: Association with early acquisition of response to thyroid hormone and estrogen. EMBO J 1990; 9:879-85.
23. O'Malley BW, McGuire WL, Kohler PO et al. Studies on the mechanism of steroid hormone regulation of synthesis of specific proteins. Recent Prog Horm Res 1969; 25:105-53.
24. Palmiter RD, Mulvihill ER, McKnight GS et al. Regulation of gene expression in chick oviduct by steroid hormones. Cold Spring Harbor Symp Quant Biol 1978; 42:639-47.
25. Schimke RT, Rhoads RE, Palacios R et al. Ovalbumin mRNA, complementary DNA and hormone regulation in chick oviduct. Karolinska Symposia on Research Methods in Reproductive Endocrinology 1973; 6:357-75.
26. Tata JR, Smith DF. Vitellogenesis: A versatile model for hormonal regulation of gene expression. Recent Prog Horm Res 1979; 35:47-95.
27. Wangh LJ. Glucocorticoids act together with estrogens and thyroid hormones in regulating the synthesis and secretion of *Xenopus* vitellogenin, serum albumin and fibrinogen. Dev Biol 1982; 89:294-8.
28. Ulisse S, Tata JR. Thyroid hormone and glucocorticoid independently regulate the expression of estrogen receptor in male *Xenopus* liver cells. Mol Cell Endocrinol 1994; 105:45-53.
29. Kelly PA, Ali S, Rozakis M. The growth hormone/prolactin receptor family. Rec Progr Horm Res 1993; 48:123-64.
30. Nicoll CS. Physiological actions of prolactin. In: Knobil E, Sawyer WH, eds. Handbook of Physiology, Section 7, Vol. 4, part 2. Washington: American Physiological Society, 1974:253-92.
31. Kikuyama S, Kawamura K, Tanaka S et al. Aspects of amphibian metamorphosis: Hormonal control. Int Rev Cytol 1993; 145:105-48.
32. Tata JR, Kawahara A, Baker BS. Prolactin inhibits both thyroid hormone-induced morphogenesis and cell death in cultured amphibian larval tissues. Dev Biol 1991; 146:72-80.
33. Denver RJ. Neuroendocrine control of amphibian metamorphosis. In: Gilbert LI, Tata JR, Atkinson BG, eds. Metamorphosis. Postem-

bryonic Reprogramming of Gene Expression in Amphibian and Insect Cells. San Diego: Academic Press, 1996:433-64.

34. Baker BS, Tata JR. Prolactin prevents the autoinduction of thyroid hormone receptor mRNAs during amphibian metamorphosis. Dev Biol 1992; 149:463-7.

35. Iwamuro S, Tata JR. Contrasting patterns of expression of thyroid hormone and retinoid X receptor genes during hormonal manipulation of *Xenopus* tadpole tail regression in culture. Mol Cell Endocrinol 1995; 113:235-43.

36. Nishikawa A, Shimizu-Nishikawa K, Miller L. Spatial, temporal and hormonal regulation of epidermal keratin expression during development of the frog *Xenopus laevis*. Dev Biol 1992; 151:145-53.

37. Miller L. Hormone-induced changes in keratin gene expression during amphibian skin metamorphosis. In: Gilbert LI, Tata JR, Atkinson BG, eds. Metamorphosis. Postembryonic Reprogramming of Gene Expression in Amphibian and Insect Cells. San Diego: Academic Press, 1996:599-624.

38. Clevenger CV, Medaglia MV. The protein tyrosine kinase p59^{fyn} is associated with prolactin (PRL) receptor and is activated by PRL stimulation of T-lymphocytes. Mol Endocrinol 1994; 8:674-81.

39. Welte T, Garimorth K, Philipp S, Doppler W. Prolactin-dependent activation of a tyrosine phosphorylated DNA binding factor in mouse mammary epithelial cells. Mol Endocrinol 1994; 8:1091-1102.

40. Ihle JN. Signaling by the cytokine receptor superfamily—just another kinase story. Trends Endocrinol Met 1994; 5:137-43.

41. Darnell JE, Kerr IM, Stark GR. Jak-STAT pathways and transcriptional activation in response to IFNs and other extracellular signaling proteins. Science 1994; 264:1415-21.

42. Daly C, Reich NC. Receptor to nucleus signaling via tyrosine phosphorylation of the P91 transcription factor. Trends Endocrinol Met 1994; 5:159-64.

43. Briscoe J, Guschin D, Rogers NC et al. JAKs, STATs and signal transduction in response to the interferons and other cytokines. Phil Trans R Soc Lond B 1996; 351:167-71.

44. Stöcklin E, Wissler M, Gouilleux F et al. Functional interactions between Stat5 and the glucocorticoid receptor. Nature 1996; 383:726.

45. Machuca I, Esslemont G, Fairclough L et al. Analysis of structure and expression of the *Xenopus* thyroid hormone receptor β (xTRβ) gene to explain its autoinduction. Mol Endocrinol 1995; 9:96-108.

46. Ulisse S, Esslemont G, Baker BS et al. Dominant-negative mutant thyroid hormone receptors prevent transcription from *Xenopus* thyroid hormone receptor β gene promoter in response to thyroid hormone in *Xenopus* tadpoles in vivo. Proc Natl Acad Sci USA 1996; 93:1205-9.

47. Wong J, Shi Y-B. Coordinated regulation of and transcriptional activation by *Xenopus* thyroid hormone and retinoid X receptors. J Biol Chem 1995; 270:18479-83.

48. Kanamori A, Brown DD. The regulation of thyroid hormone receptor β genes by thyroid hormone in *Xenopus laevis*. J Biol Chem 1992; 267:739-45.
49. Machuca I, Tata JR. Autoinduction of thyroid hormone receptor during metamorphosis is reproduced in *Xenopus* XTC-2 cells. Mol Cell Endocrinol 1992; 87:105-13.
50. Ulisse S, Iwamuro S, Tata JR. Differential responses to ligands of overexpressed thyroid hormone and retinoid X receptors in a *Xenopus* cell line and in vivo. Mol Cell Endocrinol 1997; 126:17-24.
51. Ranjan M, Wong J, Shi Y-B. Transcriptional repression of *Xenopus* TRβ gene is mediated by a thyroid hormone response element located near the start site. J Biol Chem 1994; 269:24699-705.
52. Diamond MI, Miner JN, Yoshinaga SK et al. Transcription factor interactions: Selectors of positive or negative regulation from a single DNA element. Science 1990; 249:1266-72.
53. Aronica SM, Katzenellenbogen BS. Stimulation of estrogen receptor-mediated transcription and alteration in the phosphorylation state of the rat uterine estrogen receptor by estrogen, cyclic adenosine monophosphate, and insulin-like growth factor-1. Mol Endocrin 1993; 7:743-52.
54. Cho H, Katzenellenbogen BS. Synergistic activation of estrogen receptor-mediated transcription by estradiol and protein kinase activators. Mol Endocrin 1993; 7:441-52.
55. Katzenellenbogen BS, Norman MJ. Multihormonal regulation of the progesterone receptor in MCF-7 human breast cancer cells: interrelationships among insulin/insulin-like growth factor-I, serum, and estrogen. Endocrinology 1990; 126:891-8.
56. Chen JD, Evans RM. A transcriptional co-repressor that interacts with nuclear hormone receptors. Nature 1995; 377:454-7.
57. Hörlein AJ, Näär AM, Heinzel T et al. Ligand-independent repression by the thyroid hormone receptor mediated by a nuclear receptor co-repressor. Nature 1995; 377:397-404.
58. Yen PM, Chin WW. New advances in understanding the molecular mechanisms of thyroid hormone action. Trends Endocrin Metab 1994; 5:65-72.
59. Tata JR. The action of growth and developmental hormones, evolutionary aspects. In: Goldberger RF, Yamamoto KR, eds. Biological Regulation and Development, vol. 3B, New York: Plenum Publishing. 1984:1-58.
60. Vegeto E, Cocciolo MG, Raspagliesi F et al. Regulation of progesterone receptor gene expression. Cancer Res 1990; 50:5291-5.
61. Yen PM, Liu Y, Palvimo JJ et al. Mutant and wild-type androgen receptors exhibit cross-talk on androgen-, glucocorticoid-, and progesterone-mediated transcription. Mol Endocrinol 1997; 11:162-71.
62. Hunter J, Kassani A, Winrow CJ et al. Cross-talk between the thyroid hormone and peroxisome proliferator-activated receptors in regulating peroxisome-proliferator-responsive genes. Molec Cell Endocrinol 1996; 116:213-21.

63. Nunez SB, Medin JA, Braissant O et al. Retinoid X receptor and peroxisome proliferator-activated receptor activate an estrogen responsive gene independent of the estrogen receptor. Mol Cell Endocrin 1997; 127:27-40.

64. El-Tanani MKK, Green CD. Interaction between estradiol and cAMP in the regulation of specific gene expression. Mol Cell Endocrinol 1996; 124:71-7.

65. Mora GR, Mahesh VB. Autoregulation of androgen receptor in rat ventral prostate: involvement of c-*fos* as a negative regulator. Mol Cell Endocrinol 1996; 124:111-20.

66. Hunter T. Oncoprotein networks. Cell 1997; 88:333-46.

67. Parker PJ, ed. Cell Signaling. Cold Spring Harbor: Cold Spring Harbor Press, 1996.

68. Yamauchi K, Tata JR. Purification and characterization of a cytosolic thyroid-hormone-binding protein (CTBP) in *Xenopus* liver. Eur J Biochem 1994; 225:1105-12.

69. Shi Y-B, Wong J, Puzianowska-Kuznicka M et al. Tadpole competence and tissue-specific temporal regulation of amphibian metamorphosis: roles of thyroid hormone and its receptors. Bioessays 1996; 18:391-9.

CHAPTER 8

Programmed Cell Death Is Important for Hormonal Control of Postembryonic Development

In contrast to early embryogenesis, postembryonic development is characterized by extensive cell death. The process, which is ontogenically programmed, is essential for normal development to proceed. Often an impairment of the loss of cells earmarked for elimination will prevent or disturb the differentiation and proliferation of cells programmed for further development and eventually, the acquisition of the adult phenotype. As with other processes of postembryonic development, cell death during this stage of development is strongly regulated by hormones and other signaling molecules. In considering programmed cell death or apoptosis in the context of postembryonic development it is important to distinguish it from cell death that occurs during senescence, wounding or necrosis.

Dramatic progress has been made in the last decade, thanks to the advances in molecular genetics and cell biology, in furthering our understanding of programmed cell death or apoptosis and its control. We are now beginning to have meaningful insights into the molecular steps that determine whether a cell is to die or survive. It is appropriate to begin this chapter with a brief historical account of the development of the concepts of programmed cell death leading to our present-day understanding of the process, before discussing in detail the important role played by hormones and other signaling molecules in controlling programmed cell death during postembryonic development.

Hormonal Signaling and Postembryonic Development,
by Jamshed R. Tata. © 1998 Springer-Verlag and R.G. Landes Company.

A Brief Historical Account of Programmed Cell Death and Its Genetic Regulation

The beginning of the currently enormous interest in programmed cell death (PCD) or apoptosis can be traced back to those cell morphologists and pathologists of the mid-nineteenth century who had observed that cells were derived from other cells and that changes in cellular morphology could be correlated with those of function. Among the early pioneers was Virchow who, in 1855, proposed that all diseases were due to active or passive disturbances in cell structure and function was most effective in promoting the concept of cellular basis of physiological function. Nearly forty years later, as a result of some precise microscopic observations, Schmauss described some of the most fundamental characteristics of cells undergoing the process of death, then known under such terms as 'hyperchromatosis', 'karyarthesis' and 'karyolysis'. At present, these changes in the structure of nuclei of dying cells are termed as pycnotic nuclei, chromatin condensation and nuclear breakdown. Thus, many of our present-day criteria of cell death were laid down more than a century ago. It is however in the last 20-30 years that the tools of molecular and cell biology have made it possible to begin to design, execute and verify, experimentally, some of the intricate mechanisms that underlie this process.

The next phase in the growth of our knowledge about cell death originates from observations of cellular changes following its experimental induction. For example, in 1920 Strangeways described the phenomenon of "blebbing" following irradiation of cells. This change in the structure of the cell membrane is now recognized as an important criterion of the initial steps leading to cell death. Later, in the same laboratory, Glucksmann concluded that cell death is a normal accompaniment of vertebrate ontogeny. Most importantly, he emphasized that nuclear structural changes accompanying developmental cell death should be distinguished from those caused by cellular injury or accompanying necrosis. The reader will find useful historical accounts of developmental cell death in reviews by Glucksmann[1] and Clarke and Clarke.[2]

Although in the early studies, hormones were not recognized as important external signals regulating cell death, among the first observations on cell death accompanying development were those made in hormonally controlled systems. Already in 1842, Vogt[3] had described extensive loss of cells and tissues during amphibian metamorphosis, while Berthold's classical studies in 1849[4] on loss of accessory sexual tissues during caponization of roosters is another good model for the hormonal control of cell death. In the 1960s, the availability of inhibitors of protein and RNA synthesis made it possible to demonstrate that new gene products had to be made available for cells to undergo death during development, and established the present concept of programmed cell death. This topic will be discussed below in detail (see below).

Our present-day recognition of the fact that programmed cell death is widespread and occurs normally and continuously at all stages of the life of an organism is largely due to Kerr, Wyllie and Currie.[5] By reviewing a vast number of observations on cell death occurring during development, tissue homeostasis and pathological conditions, including cancer, identified cell condensation, nuclear fragmentation and phagocytosis as major characteristic of normal cell death, and termed it apoptosis. They were also so struck by the similarity of cell death in vastly different cell types and organisms that it led them to suggest that PCD is dependent on an intrinsic death program built into the cell. What such a death program could be was not recognized until Horvitz and his colleagues, by combining techniques of genetics and molecular biology in a relatively simple organism, made a breakthrough of immense importance. By visually tracking the loss of individual cells of the nervous system of the developing nematode *Caenorhabditis elegans* in mutant and wild-type organisms, followed by gene cloning, these investigators were able to assign genes that control both cell death and cell survival.[6,7] The first gene to be thus identified, namely *ced-3*, later turned out to be homologous to a gene that encodes a mammalian cysteine protease, interleukin-1β-converting enzyme or ICE.[8] An ever-growing number of members of the *ced-3*/ICE family have now been identified in man, as well as in most vertebrates and invertebrates, many of which are implicated in PCD.[9,10] Not long after the discovery of *ced-3*, Horvitz and his colleagues described a gene, *ced-9*, which prevents cells from dying and which turned out to be homologous to the mammalian *bcl-2* gene also implicated in the process of cell survival in mammals. An ever-growing group of cell survival proteins of the *ced-9*/*bcl-2* family have now been described.[11,12] It is now clear from the immense current literature on these genes for cell death and survival that the components determining PCD have been highly conserved in evolution and that the program for cell death is expressed constitutively in all eukaryotes.

Programmed Cell Death and Postembryonic Development

The interplay between the products of cell death and survival genes, mentioned above, is summarized in Fig. 8.1. According to this scheme, a variety of external signals would modify the expression of one or more of the death effector genes or cell survival genes and thereby alter the balance between their products such that the cell would be directed either towards a death pathway or continued survival.[10,12,13] Interestingly, some genes can either act as death effector or survival genes, depending on physiological conditions or according to the cell-type in which they are expressed. For example, c-myc in the absence of serum growth factors promotes death while in their presence enhances the survival function depending on the type of cell in which it is expressed; the product of oncogene c-*rel* can also act in both capacities[14,15] An interesting property of the bcl-2 family of proteins is that they possess a dimerization domain which allows them to form homodimers or heterodimers among themselves. Whereas homodimer formation would determine which pathway the cell is to follow, the heterodimer would establish a reversible equilibrium state before either pathway becomes operational.[12,16,17] The functional consequences of this phenomenon of heterodimerization between two closely related members of the bcl-2 family, namely between bax and bcl-2, is depicted in Fig. 8.2. An uncommitted or undetermined state resulting from complexes between related or unrelated death effectors and survival promoters would be particularly important in cell dynamics during development.

Although the exact details of all the steps leading to cell death remain to be established, the outlines of the death pathway are already in place. Essentially, once the equilibrium between the opposing or conflicting effector and survival factors tip towards the death program it initiates cascades of release, activation or enhanced synthesis 'caspase' and ICE proteases which catalyze proteolysis of death substrates. Increasingly recognized as important targets for death proteases are cell matrix components. These lead to simultaneous changes in the plasma membrane ('blebbing') and of disorganization of the nuclear material, particularly manifested as chromatin condensation and chromosomal DNA degradation giving rise to the classical DNA ladder[18-20] These changes constitute the basis of several diagnostic tests for apoptosis, such as 'tunel', DNA laddering, etc. Recently, specific nucleases involved in the breakdown of chromosomal DNA breakdown have been identified.[21] The modifications

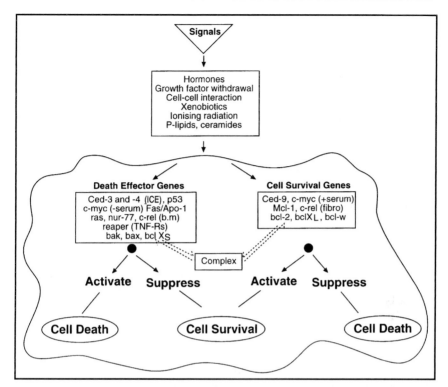

Fig. 8.1. Scheme summarizing the current concept of how the balance between the expression of cell death effector and cell survival genes determines whether a cell is to follow a death or survival pathway during PCD. Directly or indirectly the reception of one of the regulatory signals leads to the activation or repression of one or more of a few genes listed in the two boxes of death effector or survival genes. Most of these genes are expressed constitutively and the protein products of some are known to heterodimerize (see Fig. 8.2). Abbreviations: ICE, interleukin 1β activating enzyme; fibro, fibroblast; TNF-R, tumor necrosis factor receptor. The others are commonly used abbreviations for oncogenes and their products.

in plasma membrane, DNA breakdown, and possibly release of extracellular substances act as signals for macrophages to engulf and eliminate rapidly the dead cells.

According to Jacobson et al,[22] PCD serves at least five major functions during development:

1) 'Sculpting' structures, as exemplified by formation of digits by the elimination of interdigital cells, or fusion of epithelial cells prior to palate formation.
2) The removal of unwanted structures, such as the tadpole tail during metamorphosis or the elimination of Müllerian and Wolffian ducts necessary for sexual differentiation (see below).

3) Controlling numbers of cells that are overproduced, especially in the nervous system.
4) Elimination of unwanted or harmful cells, which is best illustrated by the removal of large populations of cells of the immune system.
5) Production of specialized differentiated cells, such as some skin keratinocytes, lens epithelial cells and mammalian erythrocytes. These cells are enucleated and often lack other cell organelles.

Programmed cell death is not only a major determinant of postembryonic development, but it is essential for normal development to proceed. Virtually every major late developmental process is characterized by extensive cell death, and these have been exhaustively reviewed recently.[13,15,22,23] Some of these are listed in Table 8.1. Among the best studied developmental cell death systems are the developing immune system, hematopoiesis and other stem cell differentiation pathways, developmental neuronal death, sexual dimorphic tissue regression, insect and amphibian metamorphosis and limb patterning. In the immune system, more than 95% of the thymocytes that migrate into the thymus are eliminated by active or passive cell death. This is an important function of developmental cell death since immature T cells can be harmful to the organism.[24] Among the genes involved in killing of immature thymocytes and lymphocytes are the oncogene *fas* or tumor necrosis factor (TNF-1), which initiate a proteolytic cascade leading to the activation of ICE.[25,26] Another important developmental cell death system is that relating to neural cell apoptosis.[17,27] The mechanics of the death process itself and the necessary balance between the opposing forces of death and survival (see Fig. 8.1) are similar to those encountered with PCD in other developmental system. Hematopoiesis has been intensively studied as a model of multi-stage stem cell differentiation into different adult cell-types. Multiple cytokines have been implicated, acting singly or in combination with others in the differentiation process. Interestingly, many of the same cytokines are also involved in developmental cell death.[24,28,29]

Increasing evidence that PCD is essential for normal development is now emerging from experiments based on transgenic animals and knock-out of cell death and survival genes. Gene targetting to produce *Fas*-null mice gave rise to severe lymphadenopathy and splenomegaly.[30] Further evidence that *Fas* is necessary for PCD of T lymphocytes was provided by experiments in which the lympho-

Table 8.1. Some extracellular signals inducing programmed cell death and its function during development

Cell type	Species	Signals	Induced by	Function
Neural	All invertebrates and vertebrates	Nerve and other growth factors	Withdrawal of signal	Restructuring and functional differentiation
Thymocytes, lymphocytes	All vertebrates	Fas, TNF and cytokines	Addition and withdrawal	Cell turnover and immune competence
Blood stem cells	All mammals	Multiple cytokines and hemopoietic factors	Addition and withdrawal hemopoiesis	Stem cell differentiation;
Limb bud cells	All vertebrates	Growth factors; retinoids	Addition and withdrawal	Digit formation limb development
Wolffian and Müllerian duct cells	Vertebrates	Sex hormones; growth factors	Addition	Sexual differentiation; gonadal maturation; formation of accessory sexual tissues
Digestive tract; gills; tail	Invertebrates and vertebrates	Metamorphic hormones	Addition	Tissue regression and morphogenesis

Abbreviations: TNF, tumor necrosis factor

proliferative phenotype was rescued when the Fas gene was expressed in mice carrying a 'leaky' mutation.[31] In other studies, disruption of the gene encoding bcl-x in mice was found to lead to fetal death as a result of uncontrolled cell death in some tissues during development[32] while, conversely, disruption of bax leads to a perturbation of the normal pattern of cell death during postembryonic development.[33] Clearly, programmed cell death is essential for normal development to take place.

Despite these most impressive, recent advances in the unravelling of the molecular and cellular mechanisms underlying programmed cell death, an important and intriguing question remains unanswered. Almost all the gene products required to execute the death program or to ensure cell survival are known to be expressed constitutively from very early stages of development. Yet, it

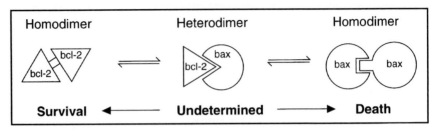

Fig. 8.2. Model to illustrate the death or survival pathway for a cell determined by direct interaction between the products of death effector and survival genes. In this example two members of the bcl family (bcl-2 and bax) heterodimerize to give an 'undetermined' state (neither death or survival). They can also exist as homodimers, in which case they will determine one or the other pathway. For details see ref. 16.

is also known for some time, that inhibitors of RNA and protein synthesis prevent PCD from proceeding (see below). This suggests that transcription and translation of some new gene products, probably induced very early after the cell death signals impinge on the targetted cell, initiate the death cascade. Except for one study, what these 'early' genes and their function are, remain a mystery. It is therefore significant that inactivation of the linked *reaper-grim* genes in *Drosophila* prevent normal cell death during development, since these genes have to be transcriptionally activated for cells to die according to a developmental program. It will be most important to identify similar genes in other species and to determine the nature and function of their protein product.[34]

Hormones and PCD

As mentioned above, Vogt's and Berthold's observations and experimentation on amphibian metamorphosis and caponization of chickens were the first examples of programmed cell death. Berthold's work on the castration of roosters and re-implantation of testis provided the first experimental evidence that internal body secretions controlled the growth and regression of specific tissues associated with sexual differentiation (cited in refs. 4,35,36). As shown in Fig. 8.3, which depicts the essence of Berthold's experiments, cell death produced by removal of the rooster's testes (withdrawal of androgen) was reversible in that replacement of androgen by transplantation of the testis caused the capon's comb and wattle to grow again. An even more interesting phenomenon of hormonal control of PCD during development is the effect of the gonads during early

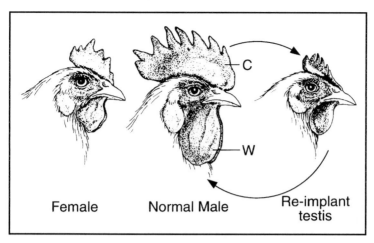

Fig. 8.3. Summary of Berthold's experiment in 1849 on tissue regression following caponization. It is the first experimental evidence for the major role hormones play in regulating cell death and morphogenesis. Castration of the rooster leads to extensive tissue regression with the loss of much of the comb and wattle (as well as other accessory sexual tissues). Replacement of testosterone, as was accomplished in Berthold's experiment by reimplantation of the testis, reverses the effects of caponization. Abbreviations: C, comb; W, wattle.

postembryonic development in the differentiation of accessory sexual accessory tissues.[37] During vertebrate development, the "gonoducts," better known as Wolffian and Müllerian ducts, undergo selective PCD under the influence of male or female sex steroids to allow only male or female accessory sexual tissues (Fig. 8.4). At some stage of fetal or postembryonic development, the Leydig cells of the testis secrete a protein hormone or growth factor called Müllerian-inhibiting substance (MIS),[38] whose action is to induce cell death and regression of the female Müllerian duct which would normally develop into the uterus or oviduct in females. At the same time, the Sertoli cells of the testis secrete androgen which promotes the differentiation of the male Wolffian duct into the seminal vesicles and vas deferens.

There are several other examples of the presence or absence of hormones regulating PCD both during development and in adults, some of these are listed in Table 8.2. Lymphocytolysis induced by glucocorticoids has been studied intensively for several years and is an ideal example of hormonally regulated PCD.[39] Useful lymphoid cell lines that are both sensitive and resistant to the hormone have proved to be useful in investigating the role of glucocorticoid receptor

Fig. 8.4. Irreversible sexual differentiation early in development caused by a death-inducing and a differentiation-promoting factor made in the testis. During postembryonic or fetal life of the sexually dimorphic organism the Leydig and Sertoli cells secrete Müllerian Inhibiting Substance (MIS) and testosterone (androgen), respectively. The effect of MIS is to induce death in the cells of the Müllerian duct, otherwise destined to

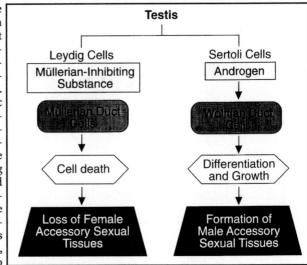

become female accessory sexual tissues. At the same time, the androgen acts on Wolffian duct cells which are induced to differentiate and grow into male accessory sexual tissues (seminal vesicles, prostate, vas deferens). The combined effect is a masculinization of the fetus during postembryonic development. Other details in ref. 37.

in the action of the hormone. In the accessory sexual tissues of mammals and mammary epithelial cells, the withdrawal of androgen or estrogen induces a marked tissue loss.[40] Induction of cell death by hormone withdrawal is the basis of much work on treatment of hormone-dependent cancers of breast, uterus and prostate with anti-estrogens and anti-androgens.[39] Interestingly, the withdrawal of the sex steroids or administration of anti-hormones leads to increased expression of some of the cell death effector genes and a lowering of that of the bcl-2-like survival genes in the hormone-dependent tumor cells. Ovarian follicle atresia is another good example of hormone mediated programmed cell death. More than 99% of follicle cells degenerate and die during reproductive life in many mammals and among the hormones implicated in follicle apoptosis are androgen and GnRH.[41] As discussed below in greater detail, perhaps the best model for studying hormonal regulation of PCD during postembryonic development is metamorphosis.

Amphibian Metamorphosis Is an Ideal Model for Studying PCD During Postembryonic Development

A major characteristic of metamorphosis in both insects and amphibia is the extensive histolysis and cell death that occurs dur-

Table 8.2. Some examples of hormonal regulation of programmed cell death and its biological function

Cells, Tissues	Species	Hormone	Addition or Withdrawal	Consequence
Thymocytes, lymphoid cells	Mammals	Glucocorticoids	Addition	Thymic involution, modified immune function
Prostate	Mammals	Androgen	Withdrawal	Involution
Uterus	Mammals	Estrogen	Withdrawal	Regression
Mammary epithelial cells	Mammals	Estrogen	Withdrawal	Involution
Larval muscle cells	Insects	Ecdysone	Addition	Muscle breakdown during metamorphosis
Larval intestinal epithelium, gills, cells of the tail,	Amphibia	Thyroid hormone	Addition	Resorption during metamorphosis
Müllerian duct	Mammals, birds	Müllerian-inhibiting substance	Addition	Tissue removal
Ovarian follicle cells	Rodents	Androgen, estrogen	Addition Withdrawal	Ovarian follicle atresia
Various larval	Insects	Juvenile hormone	Addition	Prevention of PCD during metamorphosis

ing the larval—adult transition.[42,43] The fact that all the developmental changes can be induced precociously or retarded by administering exogenous hormones, and that these could be reproduced in tissue culture renders metamorphosis an ideal model for studying programmed cell death during postembryonic development (see chapters 5,6,7). Some of the major findings from work on thyroid hormonal regulation of PCD during amphibian metamorphosis are discussed below.

Cell Death Is Extensive and Its Program Established Early in Development

Upon the onset of natural or thyroid hormone-induced metamorphosis, the amphibian tadpole undergoes rapid and substantial loss of tissue mass and cell numbers until metamorphosis is completed. As shown in Fig. 8.5, all cells of the tail (skin, muscle, nerves, connective tissue) and gills are eliminated. Extensive cell loss (50-90%) also occurs in tissues that undergo substantial remodelling,

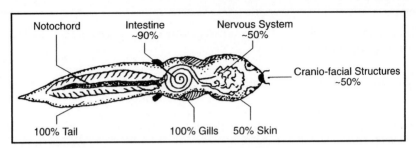

Fig. 8.5. Major tissues of the anuran tadpole undergoing extensive (50-100%) cell death during hormone-induced metamorphosis. Numbers indicate the percentage of larval cells being removed upon completion of metamorphosis.

both structurally and functionally, such as the intestine, pancreas, skin and central nervous system. Although induced by the same hormonal signal, the kinetics of cell death varies from one tissue to another and from one cell-type to another (see refs. 42,44,45,92). A particular advantage of studying PCD in the tadpole tail and intestine is that the natural in vivo process can be mimicked by adding thyroid hormone to cultures of their explants.[46-49]

Until 1968 it was not known how early in development competence is established for the larva to respond to the metamorphic stimulus of thyroid hormone. Experiments with *Xenopus* embryos and larvae then revealed that the larva acquires competence very early in development[50] As shown earlier in Fig. 5.9, *Xenopus* tadpoles at stage 45, i.e., within a week after fertilization, are capable of responding to exogenous hormone. Even at this early stage, when the tail should be growing, addition of T_3 will cause some regression of the tail. This is several weeks before endogenous TH would be secreted and natural metamorphosis begins in this species. Thus, the program for cell death during metamorphosis is laid down very early in development. The molecular basis for this early metamorphic competence resides in the expression of thyroid hormone receptors very early in development. TRα and β transcripts and proteins have been detected in early stages of *Xenopus* embryos and larvae.[51-54] In particular, high concentrations of TR mRNAs and proteins have been localized in premetamorphic *Xenopus* tadpole tail and intestine, two tissues underoing extensive or total regression. These mRNAs increase at the onset of, and during metamorphosis by a process of autoinduction (see chapter 6), thus rendering the tissue even more sensitive to the killing action of TH.

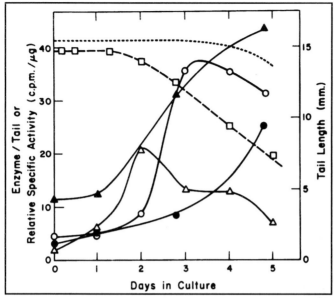

Fig. 8.6. Increase in activity of lytic enzymes and of RNA and protein synthesis accompanying the regression of *Xenopus* tadpole tails in organ culture induced by T_3. Tissue regression is expressed as diminution in tail length, which is equivalent to loss of DNA. - -, Control tail length; □ , T_3-induced tail length; ● , cathepsin; ▲ , DNase I; ○ , rate of RNA synthesis; △ , rate of protein synthesis. Based on Tata.[57]

Protein Synthesis Is Required for Initiation of PCD

Earlier studies to explain tissue regression during amphibian metamorphosis led to the conclusion that tissue lysis resulted from thyroid hormone-induced lysosomal expansion or increase in activity of latent lytic enzymes such as proteases, nucleases, phosphatases, etc. followed by macrophage infiltration.[45,55,56] However, the failure of TH to activate directly these latent enzymes or to induce cell death and tissue regression characteristic of metamorphosis by activating these enzymes with other agents meant that one had to consider the possibility of their selective synthesis under hormonal control in tissues developmentally programmed for regression.

In the mid-1960s, some laboratories were able to maintain *Xenopus* and *Rana* tadpole tails in organ cultures for sufficiently long periods, such that it was possible to induce tissue regression by the simple addition of TH to the culture medium.[55,57,58] By optimizing culture conditions, it was hence possible to induce with the hormone, several lytic enzymes such as cathepsins, RNases and DNases which

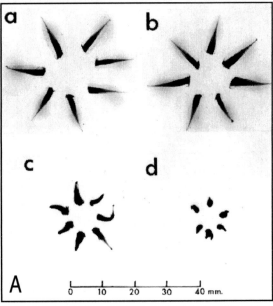

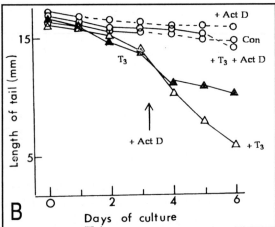

Fig. 8.7. Inhibition of RNA synthesis by actinomycin D blocks T_3-induced regression of *Xenopus* tadpole tails in organ culture. A: Photograph showing mostly muscle tissue of cultured tadpole tails at 4 and 8 days after treatment with T_3 and actinomycin D, separately and together. a: Control tails after 8 days in culture; b: after 8 days in culture with T_3 (5 x 10^{-8} M) and 5 µg actinomycin D/ml; c: after 4 days in culture with T_3 alone; d: after 8 days in culture with T_3 alone. B: Kinetics of *Xenopus* tadpole tail regression in culture following the addition of T_3 at the beginning of culture period (day 0) and actinomycin D added at day 0 or day 3. Other conditions as in A. (Adapted from ref. 57.)

were accompanied by regression of the cultured tails. Culturing also had the advantage that it was possible to simultaneously determine if the cell loss was accompanied by changes in protein and RNA synthesizing activities of the regressing tissue. As shown in Fig. 8.6, the increase in lytic enzyme activity and tissue regression following the addition of T_3 to organ cultures of *Xenopus* tadpole tails was accompanied by a burst of RNA and protein synthesizing activity. This finding raised two important questions. Is the enhanced protein synthetic activity necessary for cell death? Are any new proteins synthesized at the onset of tissue loss, and if so, what is their nature?

In the 1960s, specific inhibitors of RNA and protein synthesis became available, such as actinomycin D, puromycin and cycloheximide. When these inhibitors were added to organ cultures of tadpole tails, along with or after T_3, a paradoxical result was obtained. These cytotoxic agents, which normally kill cells, were found to protect cells programmed to die during postembryonic development. The action of actinomycin D, illustrated in Fig. 8.7A, quite clearly shows that the size of the tail muscle (seen as dark image, the surrounding skin and tail fin not being visible) is maintained at almost normal values until 8 days after starting the cultures. When the kinetics of tail regression were measured (Fig. 8.7B), actinomycin D not only blocked tail regression when added along with T_3 at the beginning of the culture, but also when administered after T_3 had initiated regression (DNA measurements show that measuring tail length is a good index of cell loss). Thus, ongoing protein synthesis is necessary for PCD to be initiated, as well as for it to continue. Similar results with inhibitors of protein synthesis were also obtained by Weber.[59] Later, Lockshin's laboratory and others also found the same to be true for cell death in muscle and other tissues in ecdysteroid-regulated metamorphosis in insects.[43,60] The requirement for protein synthesis for normal PCD has also been demonstrated for limb differentiation and digitization in chick postembryonic development which indicates this requirement to be of general occurrence.[61]

As to the nature of any new proteins whose de novo or enhanced synthesis is essential for developmental PCD, early experiments based on radioactive proteins proved to be fruitless. Attempts to identify new proteins by double and triple isotope labeling could only indicate that any such protein made would be present in the regressing tadpole tail cultures would constitute only 0.01% of all new proteins synthesized.[58] Later, using high resolution, two-dimensional gel electrophoresis, other investigators[60] did find some new proteins in degenerating insect muscle during metamorphosis, but their identity remains unknown. It is only when recombinant DNA techniques, such as subtractive hybridization and reverse transcriptase-polymerase chain reaction (RT-PCR) were applied that it became possible to obtain some idea about the qualitative and quantitative aspects of new genes expressed during PCD in amphibian metamorphosis.

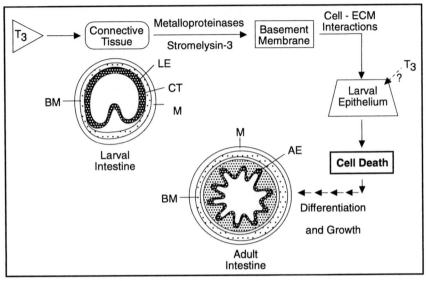

Fig. 8.8. Scheme emphasizing the importance of activation of metalloproteinases and cell-extracellular matrix (ECM) interactions in T_3-induced cell death followed by differentiation and remodelling of the anuran larval small intestine. The larval intestine is a relatively simple structure with one loop, as shown in the cross-section diagram, known as the typhlosole. The major target of T_3 is the connective tissue which secretes the metalloproteinase stromelysin-3. The enzyme acts mainly on the basement membrane and modifies cell-ECM interactions which, in turn, lead to the loss of 85-90% of the larval epithelial cells. The surviving cells undergo differentiation to the adult-type, convoluted epithelium with substantial thickening of the basement membrane. Cell death is necessary for later differentiation and growth. Abbreviations: LE, larval epithelium; M, muscle; BM, basement membrane; CT, connective tissue; AE, adult epithelium. Further details can be found in refs. 64,70,72.

Involvement of Cell-Cell Interaction and Extracellular Matrix in PCD

Another major tissue undergoing extensive cell death during vertebrate and invertebrate metamorphosis is the gut or small intestine. In *Xenopus*, nearly 90% of the small intestinal epithelium is lost following the onset of metamorphosis.[62,63] But unlike tissues undergoing total regression, such as the tail and gills, the remaining 10% of the cells undergo rapid differentiation and multiplication to generate the adult digestive tract, also under the control of thyroid hormone. What determines which cells are earmarked for death, and which for survival and differentiation under the control of the same hormonal signal? The answer to this intriguing question remains to be established. Meanwhile, Ishizuya-Oka and her colleagues have carried out detailed studies on the regression and remodelling of the amphibian intestine during metamorphosis.[49,64,65] Because the structure of the larval intestine is less complex than that of the tail, it has been easier to study the involvement of cell-cell interactions in the cell death process. At the same time, since a new structure (adult) replaces the larval structure, the intestine also offers a model for analyzing the involvement of cell-extracellular matrix (ECM) in morphogenesis. Recent studies on morphological and functional changes in a variety of developmental and pathological systems have clearly established the importance of basement membrane, ECM, matrix metalloproteinases, cell adhesion molecules, integrins, etc. in the regulation of PCD.[66-69]

By combining detailed analysis of ultrastructure, immunocytochemistry, and molecular biology, the laboratories of Ishizuya-Oka and Shi have begun to piece together several important elements of the cell matrix and basement membrane that are important in T_3-induced cell death in the small intestine of *Xenopus* larvae, both in vivo and in organ culture.[64,65,70-72] According to the simplified scheme shown in Fig. 8.8, the major target of T_3 is the connective tissue lying between the very simple larval intestinal epithelium and basement membrane. The hormone is thought to induce metalloproteinases which would act on the basement membrane and thus facilitate cell-ECM interaction. In situ hybridization has allowed the localization of the transcripts of one of the metalloproteinases, stromelysin-3, which is transiently expressed during metamorphosis, in the connective tissue.[70] The cell-ECM interactions induce PCD of the larval epithelium, as a result of which multiple differentiative steps lead to the formation of the adult intestine, characterized by a substantial thickening of the basement membrane and the formation of a new, highly convoluted intestinal epithelium. Co-culture experiments with different regions of the larval intestine have established that the death of larval epithelial cells is essential for differentiation and development of the adult intestine.

Besides the intestine, several other tissues and organs undergo major structural reorganization during metamorphosis in amphibia and insects. Particularly, well studied are the transitions from larval to adult skin or cuticle and brain[43,73-75] (see chapter 5). These changes are important since they accompany the acquisition of important functional characteristics. Not only are they critical for the metamorphosis of free-living larvae but similar cell-cell and cell-ECM interactions also pay a central role during postembryonic development of mammals.[66,67,69]

Other Hormones Potentiate and Inhibit T_3-induced PCD

Exogenous glucocorticoids enhance T_3-induced amphibian metamorphosis and prolactin (PRL) suppresses this process (chapters 5,7), including the regression of the tadpole tail. In organ cultures of the tails, these effects are paralleled by the up- or down-regulation of the expression of TR and RXR. How these receptor mRNA changes correlate with T_3-induced PCD is summarized in Fig. 8.9. In this experiment the synthetic glucocorticoid dexamethasone (Dex) did not induce apoptosis on its own, but speeded up tail regression induced by T_3. PRL slowed down considerably the Dex-accelerated rate of cell death as it did with T_3 alone. We have seen earlier that these responses were paralleled by the effects of Dex and PRL on the autoinduction of TRβ mRNA (see chapter 7). This hormonal interplay is also evident when one expresses the data shown in Fig. 8.9 in terms of the expression of genes encoding cell death effector genes (see below).

In Fig. 8.9, measurements of tail length (or the DNA content) represent the sum of all the apoptotic effects of T_3 on the whole tail. However, unlike the intestine, this organ is highly complex and made up of widely different cell types (e.g., epidermis, muscle, nerve, vascular tissue, etc.). The question then arises as to whether all types of cells are killed at the same rate and in the same way by T_3. Histological examination of *Xenopus* tadpole tails and intestine undergoing regression during natural and hormone-induced metamorphosis, clearly reveals that different cell-types do not display the same sensitivity and kinetics of cell death in response to the hormone. In the intestine the transient expression of stromelysin-3, a metalloproteinase associated with apoptosis is preferentially expressed in the connective tissue at the onset of regression, which is compatible with the modification of the basement membrane and epithelial transformation that follow (see refs. 64,70; Fig. 8.8). In the tadpole tail it is the collagen-rich fin surrounding the body of the tail which disappears well before the other underlying tissues. As can be seen in the histological sections of the tail hormonally induced to regress in vivo, not all tissues are degraded simultaneously (Fig. 8.10).[76] There appears to be a spatial and temporal hierarchy, whereby the progression of tail regression exhibits the sequence of tissue degeneration which proceeds in the order: fin—skin—notochord—muscle—spinal cord and other tissues. One also notices a posterior-anterior direction of regression, i.e., regression begins at

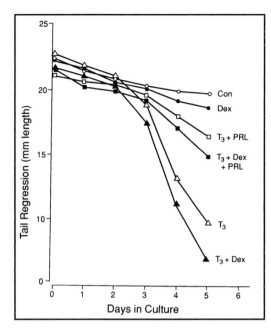

Fig. 8.9. Opposing effects of prolactin (PRL) and the synthetic glucocorticoid dexamethasone (Dex) on the kinetics of T_3-induced premetamorphic *Xenopus* tadpole tail regression in organ culture. Regression measured as the reduction in tail length is roughly equivalent to loss of cells measured as DNA content. Concentrations of hormones: T_3, 2 x 10^{-9} M; Dex, 10^{-7} M; PRL, 0.25 i.u./ml. Other details in ref. 84.

the distal end of the tail (tail tip) and proceeds towards the main body of the tadpole. Such findings are compatible with the well known fact that, just as in morphogenesis, apoptosis accompanying tissue and organ regression also follows a well-defined spatio-temporal pattern during development.

Early and 'Direct-Response' Genes Associated with PCD

In following the temporal pattern of cell death responses during PCD, it is desirable to identify genes that are expressed early following the death signal. Although it does not automatically follow that the genes expressed early play a causal role with respect to later events, the characterization of such genes could help in understanding the significance of later responses. At the same time, if any of the early genes also behave as 'direct-response' genes, i.e., they are activated by the death-inducing signal, then it strengthens the case for those genes to be placed at the top of a cascade of cell death inducing processes. The laboratories of Brown and Shi have attempted to characterize by techniques of subtractive hybridization as many genes as possible that are up or down-regulated in regressing *Xenopus* tail and intestine, and then compare the pattern of their expression in tissues programmed for morphogenesis and growth (particularly the developing limb).[77-80]

Fig. 8.10. Different tissues of the *Xenopus* tadpole tail are sensitive to the action of T_3 to different extents. The figure shows transverse sections of tails of stage 54/55 *Xenopus* tadpoles maintained for 7 days in culture following the addition of T_3 and prolactin (PRL). A: Control; B: T_3-treated; C: PRL added 3 days after culture with T_3; D: PRL added 1 day after culture with T_3. Abbreviations: DTF and VTF, dorsal and ventral fins, respectively; SC, spinal cord; N, notochord; M, muscle; VT, connective tissue; BV, blood vessel. Other details in ref. 76.

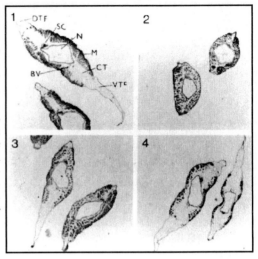

Interestingly, most of the genes whose expression is altered during tail regression are upregulated (see Table 8.3). These can be divided into four groups:

1) Those encoding transcription factors which includes TRβ.
2) Proteinases including metalloproteinases which are implicated in cell-ECM interactions.
3) ECM proteins themselves such as fibronectin and integrin, but these are likely to be indirectly upregulated.
4) Others, among which is an unidentified gene specifically expressed in the tail and is conspicuous by the fact that most genes listed in Table 8.3 are ubiquitously expressed and also upregulated in non-regressing tissues, such as those of the limbs.

An interesting upregulated gene is Type III deiodinase which converts L-thyroxine (T_4) to the active thyroid hormone 3,3',5-triiodo-L-thyronine (T_3) by removal of the iodine atom in the 5' position. A simple-minded view would be that since T_4 is the abundant precursor of T_3, which is the active hormone, the activation of this gene would ensure a high level of intracellular T_3. Another upregulated gene in a tissue predominantly programmed for regression during the early stages of metamorphosis, namely the intestine, is like sonic the hedgehog.[81] This gene, which has been characterized as encoding an important morphogen during early embryogenesis, was found to be maximally upregulated at meta-

Table 8.3. Genes upregulated in Xenopus tadpole tail during metamorphosis

Transcription factors	Proteinases	ECM	Others
TRβ(DR)	Stromelysin-3(DR)	Fibronectin	Type III iodothyronine, deiodinase(DR)
Zn finger(BTEB)(DR)	Collagenase 3	Integrin α-1	Unidentified-1(DR)
bZip(E4BP4)	FAPα		Unidentified-2(DR)
bZip(Fra-2)	N-Aspartyl dipeptidase		CRF-binding protein
			Unidentified-3(TS)

(DR), Direct response gene; (TS), tail-specific. Adapted from ref. 79 which should be consulted for further details.

morphic climax (stage 62) during both natural and TH-induced metamorphosis of *Xenopus* tadpoles. The significance of its upregulation during tissue regression is unclear since it was also upregulated in a growing tissue like the limb. There are also some direct-response upregulated genes which have not yet been identified. Among the few downregulated genes during tissue regression is that encoding for Intestinal Fatty Acid Binding Protein.[82] Although this gene is expressed very early during *Xenopus* development, it is transiently downregulated at metamorphic climax, although this process is not thought to be due to a direct response to TH.

As has been discussed earlier (chapters 6,7), the most rapidly upregulated of the direct-response genes during metamorphosis is TRβ gene. The rapidity of the response is particularly well established in cell and organ cultures, as in vivo (see Fig. 6.7). TRβ upregulation has been studied in our laboratory in detail in organ cultures of pre-metamorphic tadpole tails in which regression is induced by T_3.[83-85] The T_3-induced regression of tails in culture is potentiated by dexamethasone (see Fig. 8.9) as are all the other processes of metamorphosis. That the effect of the hormones in organ culture is physiological is suggested by the fact that the dose-response pattern follows that in vivo. However, as can be seen in Fig. 8.11, at both sub-optimal and maximum doses the responses of TRβ and RXRγ genes are sharply in contrast. The synthetic glucocorticoid has no effect on TRβ mRNA accumulation, whereas it upregulates RXRγ (as it does RXRα and β) mRNA. What Dex also does is to restore the 50% downregulation by T_3 of RXRγ mRNA to values 50% above the normal. In confirmation of a result described in the preceding chapter (Figs. 7.9, 7.10), the combined effect of T_3 and Dex would be to upregulate the expression of both members of functionally relevant TR-RXR heterodimer.

In an attempt to clarify the role of rapid upregulation by T_3 of TRβ gene in PCD, Ulisse et al[85] have taken the dominant-negative receptor approach (see chapter 6; Fig. 6.12). They first showed that in XTC-2 cells transfected with both natural and artificial dominant negative mutant TRβ the auto-upregulation of TRβ was fully abolished (see Fig. 6.12). There remained to be resolved the question of whether this phenomenon could also be reproduced in a living tissue of the tadpole. To do this they devised a technique based on the microinjection of DNA constructs into Xenopus tadpole tail muscle.[86] In this way they were able to show that in pre-metamorphic Xenopus tadpole tails injected with various expression vectors and Xenopus and human dominant-negative TRβ constructs the autoinduction of TRβ accompanying the T_3-induced tail regression was totally repressed (see Fig. 8.12). The next step is now to test the possibility using a combination of molecular analysis and in situ hybridization techniques, that inhibition of TRβ autoinduction will actually prevent the onset of PCD and tissue regression. If that turns out to be the case, then it would strongly support the suggestion made in chapters 6 and 7 that autoinduction of receptor genes is important for all developmental actions of the hormone.

In a recent paper from Brown's laboratory,[87] it has been argued that it is important to study downregulated genes as a group so that these may throw some light on genes needed for larval cell survival or are not essential in adult life. Using a technique of gene screening, they were able to identify four genes encoding extracellular glycoproteins expressed in the tadpole tail. By using a combination of in situ hybridization and immunocytochemistry, these glycoproteins were found to be expressed in apical cells of the epidermis throughout the tadpole. The epidermis is well-known to undergo structural modifications during metamorphosis. What is particularly interesting is that the expression of these four genes ceases in all epidermal apical cells and not only in tissues programmed for death. This finding raises the question as to whether a similar downregulation takes place for other cell survival genes during cell death.

Genes Controlling Cell Death and Cell Survival

During development, cell survival and death are currently thought to be dictated by a balance between the expression of survival and death genes, such as those listed in Fig. 8.1 (see above).[10,12,13,15] Since virtually all of these genes have been found to be expressed during development, as well as in adult organisms, one could envisage three possible ways in which they are regulated during developmental cell death:

1) That one or more survival gene is downregulated or shut down.
2) That one or more cell death effector genes is upregulated or 'switched on.'
3) That both 1) and 2) occur, but an unknown factor is also involved.

In our laboratory we investigated the first possibility that the survival genes may be differentially regulated in tissues programmed for growth and regression during Xenopus metamorphosis.[88] For this purpose, Xenopus *bcl-2* like genes were cloned, and two members of this family (xR11 and xR1), closely related to bcl-X_L, were studied in detail, as it is considered to be a major gene conferring survival function on cells.[89] Both xR11 and xR1 exhibited all the criteria of protecting heterologous cells against the apoptotic action of cytotoxic agents, such as staurosporine and cycloheximide, or cell death induction by c-Myc. However, both xR1 and xR11 continued to be expressed in regressing as well as non-regressing tissues during natural and TH-induced metamorphosis. In this respect, it is significant that no cell survival gene has been shown to be repressed or substantially downregulated during PCD accompanying development.[12,15]

The above negative result led us to examine the expression of some cell death effector genes during tadpole tail regression in culture. These included stromelysin-3, which has been shown above to be activated during tadpole intestinal resorption,[70,71] nur-77, a member of the steroid nuclear receptor family, and ICE which are also widely expressed during cell death.[15] The genes encoding these cell death effectors are constitutively expressed in pre-metamorphic tadpole tails, as in all tissues. But as summarized in Fig. 8.13, their mRNAs were upregulated relatively moderately during T_3-induced tail regression. What is significant about these results is that prolactin and dexamethasone, which inhibit and potentiate T_3-induced cell death, respectively, had no effect on their own on the levels of these mRNAs. However, PRL suppressed the activation by T_3 of both mRNAs, which is compatible with the physiological interplay among these three

Fig. 8.11. Dose-dependent and contrasting effects of T$_3$ (T) and dexamethasone (Dex) on steady-state levels of TRβ and RXRγ in *Xenopus* tadpole tails in culture. The relative mRNA concentrations of hormone-treated tails are compared with those of controls, after 48 hours of culture. Concentrations of hormone used are indicated on the figure. Other details in ref. 84.

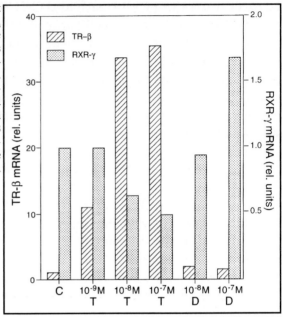

Fig. 8.12. Dominant-negative *Xenopus* and human TRβ mutants (mt-xTRβ and △-hTRβ) block the activation by T$_3$ of thyroid response element (TRE) when injected into *Xenopus* tadpole tail muscle in vivo. Tadpole tails were injected with different chloramphenicol acetyl transferase (CAT) constructs (A: p(-200/+87) xTRβ-CAT; B: F2-tk-CAT; C: TREpal-tk-CAT) with or without mt-xTRβ or △-hTRβ1. The following day the tails were excised and cultured for 5 days in the presence (hatched bars) or absence (open bars) of 10 nM T$_3$, after which CAT activity was measured. See ref. 85 for more details.

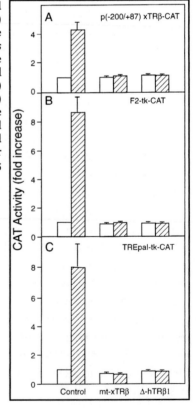

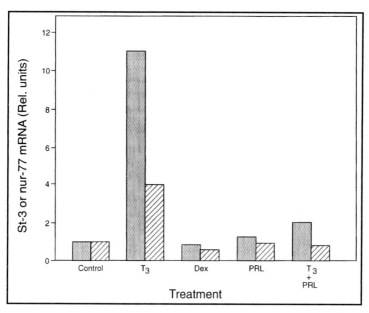

Fig. 8.13. T_3 activates expression of stromelysin-3 (St-3) and nur-77 in *Xenopus* tadpole tail in cultures. The steady-state levels of the mRNAs in tadpole tails, treated with or without (Control) hormones, exactly as in the experiment of Fig. 8.11, were determined after 48 hours of culture. Concentrations of hormones: T_3, 10 nM; Dex, 0.1 μM; PRL (prolactin), 0.25 i.u./ml.

hormones in metamorphosis (see chapter 7). Recently a cell-line (XLT-15) has been derived from myoblast cells of the Xenopus tadpole tail which undergoes apoptosis in response to the addition of T_3.[90] During this process an ICE-like cell death effector (CPP32/apopain/Yama) gene was activated. This cell-line should thus make it possible to carry out analysis of various aspects of hormonal regulation of PCD in greater depth. Similar results of a selective hormonal induction of cell death effectors have recently been obtained with T_3 alone in Demeneix's laboratory[91] with the expression of *bax*, a well-established cell death effector gene.[12,16,33] If these modest increments in the steady-state levels of cell death effector mRNAs are reflected in that of their protein products, and that these increments are physiologically meaningful, then these results suggest that small activations of cell death effector genes, without any decline in the expression of survival genes, may be sufficient to tip the balance towards cell death.

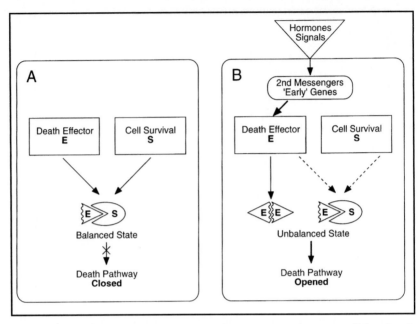

Fig. 8.14. A working model suggesting how hormones or other extracellular signals would activate programmed cell death during postembryonic development. A. It is proposed that in the absence or at sub-threshold levels of a hormone the products of cell death effector (E) and cell survival (S) genes would be in a balanced state so that PCD would not ensue. B. When the hormone or a death-inducing signal reaches its target cell, the second messengers or 'early' gene products would activate cell death effector genes but not suppress the survival genes. The resulting imbalance between E and S would allow the cell death pathway to be opened, thus setting into motion the mechanisms underlying apoptosis. The model is based on many assumptions and few definitive results, but it can be experimentally verified. See also Figs. 8.1 and 8.2.

If one accepts the above assumptions for the time being, namely that programmed cell death during postembryonic development is determined by the balance between physically interacting products of death effector and survival genes, and that hormones and other extracellular signals that induce PCD do not suppress cell survival genes but activate death effector genes, then it is useful to consider the simple working model proposed in Fig. 8.14. According to this scheme, the cells of a developing tissue would be in an undetermined or uncommitted state in the absence of a hormonal or other signal that induces cell death. In this state, a balance would be achieved between the products of one or more death effector and cell survival genes. This balance or equilibrium may be due to some physical interaction or heterodimerization of the two opposing factors (see

Figure 8.2). When at an appropriate developmental stage the hormonal signal interacts with its receptor and via second messengers or 'early' gene products, activates death effector genes, but does not affect cell survival genes, the undetermined state is perturbed due to an excess of the death effector products. As a consequence, the latter can form homodimers or interact with other constituents not yet discovered, the net effect being to lead the cell into a death pathway. This hypothesis is based on many assumptions and a few early results based on the hormonal regulation of programmed cell death during amphibian metamorphosis. However, the model can be experimentally tested and only further work can establish the validity of this suggestion as a general principle underlying developmental programmed cell death.

References

1. Glucksmann A. Cell deaths in normal vertebrate ontogeny. Biol Rev 1951; 26:5984-6.
2. Clarke PGH, Clarke S. Nineteenth century research on naturally occurring cell death and related phenomena. Anat Embryol 1996; 193:81-99.
3. Vogt C. Untersuchungen über die Entwicklungsgeschichte der Geburtshelferkroete *(Alytes obstetricans)*. 1842.
4. Gorbman A, Bern H A. A Textbook of Comparative Endocrinology. New York: John Wiley, 1962.
5. Kerr JFR, Wyllie AH, Currie AR. Apoptosis: a basic biological phenomenon with wide-ranging implication in tissue kinetics. Br J Cancer 1972; 26:239-57.
6. Horvitz H, Ellis H, Sternberg P. Programmed cell death in nematode development. Neurosci Comment 1982; 1-56-65.
7. Ellis HM, Horvitz HR. Genetic control of programmed cell death in the nematode C. elegans. Cell 1986; 44:817-29.
8. Yuan J, Shaham S, Ledoux S et al. The C. elegans cell death gene ced-3 encodes a protein similar to mammalian interleukin-1 β-converting enzyme. Cell 1993; 75:641-52.
9. Steller H. Mechanisms and genes of cellular suicide. Science 1995; 267:1445-9.
10. Chinnaiyan A, Dixit V. The cell-death machine. Curr Biol 1996; 6:555-62.
11. Hengartner MO, Horvitz HR. C. elegans cell survival gene ced-9 encodes a functional homolog of the mammalian proto-oncogene bcl-2. Cell 1994; 76:665-76.
12. Korsmeyer S. Regulators of cell death. Trends Genet 1995; 11:101-5.
13. Ellis RE, Yuan JY, Horvitz HR. Mechanisms and functions of cell death. Ann Rev Cell Biol 1991; 7:663-98.

14. Evan G, Harrington E, Fanidi A et al. Integrated control of cell proliferation and cell death by the c-*myc* oncogene. Phil Trans R Soc B 1994; 345:269-75.
15. Dexter TM, Raff MC, Wyllie AH, eds. Death from inside out: the role of apoptosis in development, tissue homeostasis and malignancy. Phil Trans R Soc B 1994; 231-3.
16. Oltvai ZO, Milliman CL, Korsmeyer SJ. Bcl-2 heterodimerizes in vivo with a conserved homolog, Bax, that accelerates programed cell death. Cell 1993; 74:609-19.
17. Dibenedetto AJ, Pittman RN. Death in the balance. Persp Dev Neurobiol 1996; 3:111-20.
18. Kuchino Y, Müller WEG, eds. Apoptosis. Berlin: Springer, 1996.
19. Tomei LD, Cope FO, eds. Apoptosis: The Molecular Basis of Cell Death. Cold Spring Harbor, New York: Cold Spring Harbor Laboratory Press, 1991.
20. Lavin M, Watters D. Programmed Cell Death. The Cellular and Molecular Biology of Apoptosis. Harwood: Chur, 1993.
21. Tamima S, Shiokawa D. An endonuclease responsible for apoptosis. In Kuchino Y, Müller WEG, eds. Progr Mol Subcell Biol 1996; Berlin: Springer 16:1-12.
22. Jacobson MD, Weil M, Raff MC. Programmed cell death in animal development. Cell 1997; 88:347-54.
23. White E. Life, death, and the pursuit of apoptosis. Gene Dev 1996; 10:1-15.
24. Nagata S. Apoptosis by death factor. Cell 1997; 88:355-65.
25. Wallach D. Cell death induction by TNF: A matter of self control. Trends Biochem Sci 1997; 22:107-9.
26. Martin SJ, Green DR. Protease activation and apoptosis: death by a thousand cuts. Cell 1995; 82:349-52.
27. Bredesen DE. Genetic control of neural cell apoptosis. Persp Dev Neurobiol 1996; 3:101-9.
28. Sachs L, Lotem J. Apoptosis in the hemopoietic system. In: Gregory CD, ed. Apoptosis and the Immune Response. Wiley-Liss, 1995:371-403.
29. Gregory CD, ed. Apoptosis and the Immune Response. New York: Wiley-Liss, 1995.
30. Adachi M, Suematsu S, Konda T et al. Targeted mutation in the *Fas* gene causes hyperplasia in the peripheral lymphoid organs and liver. Nature Genet 1995; 11:294-300.
31. Wu J, Zhou T, Zhang J et al. Correction of accelerated autoimmune disease by early replacement of the mutated *Ipr* gene with the normal *Fas* apoptosis gene in the T cells of transgenic MRL-*Ipr/Ipr* mice. Proc Natl Acad Sci USA 1994; 91:2344-8.
32. Motoyama N, Wang F, Roth KA et al. Massive cell death of immature hematopoietic cells and neurons in Bcl-x-deficient mice. Science 1995; 267:1506-10.

33. Deckwerth TL, Elliott JL, Knudsen CM et al. BAX is required for neuronal death after trophic factor deprivation and during development. Neuron 1996; 17:401-11.
34. White K, Grether ME, Abrams JM et al. Genetic control of programmed cell death in Drosophila. Science 1994; 264:677-83.
35. Turner CD, Bagnara JT. General Endocrinology. Philadelphia: Saunders, 1971.
36. Baulieu E-E, Kelly PA, eds. Hormones. From Molecules to Disease. Paris: Hermann, 1990.
37. Jost A. Hormonal control of the masculinization of the body. In: Baulieu E-E, Kelly PA, eds. Hormones. Paris: Hermann, 1990:439-42.
38. Donahoe PK, Budzik GP, Treistad R et al. Müllerian-inhibiting substance: an update. Rec Prog Hormone Res 1982; 38:279-326.
39. Thompson EB. Apoptosis and steroid hormones. Mol Endocrinol 1994; 8:665-73.
40. Tenniswood MP, Guenette RS, Lakins J et al. Active cell death in hormone-dependent tissues. Canc Metast 1992; 11:197-220.
41. Hsueh AJW, Billig H, Tsafriri A. Ovarian follicle atresia: A hormonally controlled apoptotic process. Endocrin Rev 1994; 15:707-21.
42. Beckingham Smith K, Tata JR. The hormonal control of amphibian metamorphosis. In: Graham C, Wareing PF, eds. Developmental Biology of Plants and Animals. Oxford: Blackwell, 1976:232-45.
43. Lockshin RA. Cell death in metamorphosis. In: Bowen ID, Lockshin RA, eds. Cell Death in Biology and Pathology. London: Chapman and Hall, 1981:79-121.
44. Gilbert LI, Frieden E, eds. Metamorphosis: A Problem in Developmental Biology. New York: Plenum Press, 1981.
45. Yoshizato K. Biochemistry and cell biology of amphibian metamorphosis with a special emphasis on the mechanism of removal of larval organs. Int Rev Cytol 1989; 119:97-149.
46. Tata JR. Gene expression during metamorphosis: An ideal model for post-embryonic development. BioEssays 1993; 15:239-48.
47. Tata JR. Hormonal regulation of programmed cell death during amphibian metamorphosis. Biochem Cell Biol 1994; 72:581-8.
48. Tata JR. Hormonal signaling and amphibian metamorphosis. Adv Dev Biol 1997; 5:237-74.
49. Ishizuya-Oka A, Shimozawa A. Induction of metamorphosis by thyroid hormone in anuran small intestine cultured organotypically in vitro. In Vitro Cell Dev Biol 1991; 27A:853-57.
50. Tata JR. Early metamorphic competence of *Xenopus* larvae. Dev Biol 1968; 18:415-40.
51. Yaoita Y, Brown DD. A correlation of thyroid hormone receptor gene expression with amphibian metamorphosis. Genes Dev 1990; 4:1917-24.

52. Kawahara A, Baker B, Tata JR. Developmental and regional expression of thyroid hormone receptor genes during *Xenopus* metamorphosis. Development 1991; 112:933-43.
53. Eliceiri BP, Brown DD. Quantitation of endogenous thyroid hormone receptors α and β during embryogenesis and metamorphosis in *Xenopus laevis*. J Biol Chem 1994; 269:24459-65.
54. Fairclough L, Tata JR. An immunocytochemical analysis of expression of thyroid hormone receptor α and β proteins during natural and thyroid hormone-induced metamorphosis in *Xenopus*. Dev Growth Differn 1997; 39:273-83.
55. Weber R. Tissue involution and lysosomal enzymes during anuran metamorphosis. In: Dingle JT, Fell HB, eds. Lysosomes in Biology and Pathology. Vol. I. Amsterdam: North-Holland, 1969:437-61.
56. Atkinson BG. Biological basis of tissue regression and synthesis. In: Gilbert LI, Frieden E, eds. Metamorphosis. A Problem in Developmental Biology. New York: Plenum Press, 1981:397-444.
57. Tata JR. Requirement for RNA protein synthesis for induced regression of the tadpole tail in organ culture. Dev Biol 1966; 13:77-94.
58. Beckingham Smith K, Tata JR. Cell death. Are new proteins synthesised during hormone-induced tadpole tail regression? Exp Cell Res 1976; 100:129-46.
59. Weber R. Inhibitory effect of actinomycin on tail atrophy in *Xenopus* larvae at metamorphosis. Experientia 1965; 21:665-6.
60. Wadewitz AG, Lockshin RA. Programmed cell death: dying cells synthesize a co-ordinated, unique set of proteins in two different episodes of cell death. FEBS Lett 1988; 241:19-23.
61. Saunders JW Jr. Death in embryonic systems. Science 1966; 154:604-12.
62. McAvoy VW, Dixon KE. Cell proliferation and renewal in the small intestinal epithelium of metamorphosing and adult *Xenopus laevis*. J Exp Zool 1977; 202:129-38.
63. Hourdry J, Dauca M. Cytological and cytochemical changes in the intestinal epithelium during anuran metamorphosis. Int Rev Cytol (Suppl) 1977; 5:337-85.
64. Ishizuya-Oka A. Apoptosis of larval cells during amphibian metamorphosis. Microsc Res Tech 1996; 34:228-35.
65. Ishizuya-Oka A, Shimozawa A. Inductive action of epithelium on differentiation of intestinal connective tissue of *Xenopus laevis* tadpoles during metamorphosis in vitro. Cell Tissue Res 1994; 277:427-36.
66. Gospodarowicz D, Fujii DK, Giguere L et al. The role of the basal lamina in cell attachment, proliferation, and differentiation. Tumor cells vs normal cells. In: Murphy GP, Sandberg AA, Karr JP, eds. Progress in Clinical and Biological Research, vol. 75A. New York: Liss, 1981:95-132.
67. Trelstad RI, ed. The Role of Extracellular Matrix in Development. New York: Liss, 1984.

68. Salamonsen LA. Matrix metalloproteinases and their tissue inhibitors in endocrinology. Trends Endocrinol Metab 1996; 7:28-34.
69. Meredith Jr JE, Schwartz MA. Integrins, adhesion and apoptosis. Trends Cell Biol 1997; 7:146-50.
70. Ishizuya-Oka A, Ueda S, Shi Y-B. Transient expression of stromelysin-3 mRNA in the amphibian small intestine during metamorphosis. Cell Tissue Res 1996; 283:325-9.
71. Patterton D, Hayes WP, Shi Y-B. Transcriptional activation of the matrix metalloproteinase gene *stromelysin-3* coincides with thyroid hormone-induced cell death during frog metamorphosis. Dev Biol 1995; 167:252-62.
72. Stolow MA, Ishizuya-Oka A, Su Y, Shi Y-B. Gene regulation by thyroid hormone during amphibian metamorphosis: implications on the role of cell-cell and cell-extracellular matrix interactions. Am Zool 1997; in press.
73. Fox H. Changes in amphibian skin during larval development and metamorphosis. In: Balls M, Bownes M. Metamorphosis. Oxford: Clarendon Press, 1985:59-87.
74. Truman JW, Levine RB, Weeks JC. Reorganization of the nervous system during metamorphosis of the moth, *Manduca sexta*. In: Balls M, Bownes M. Metamorphosis. Oxford: Clarendon Press, 1985:127-44.
75. Yoshizato K. Cell death and histolysis in amphibian tail during metamorphosis. In: Gilbert LI, Tata JR, Atkinson BG, eds. Metamorphosis. Postembryonic Reprogramming of Gene Expression in Amphibian and Insect Cells. San Diego: Academic Press, 1996:647-71.
76. Tata JR, Kawahara A, Baker BS. Prolactin inhibits both thyroid hormone-induced morphogenesis and cell death in cultured amphibian larval tissues. Dev Biol 1991; 146:72-80.
77. Shi Y-B, Brown DD. The earliest changes in gene expression in tadpole intestine induced by thyroid hormone. J Biol Chem 1993; 268:20,312-17.
78. Wang Z, Brown DD. Thyroid hormone-induced gene expression program for amphibian tail resorption. J Biol Chem 1993; 268:16,270-8.
79. Brown DD, Wang Z, Furlow JD et al. The thyroid hormone-induced tail resorption program during *Xenopus laevis* metamorphosis. Proc Natl Acad Sci USA 1996; 93:1924-29.
80. Shi Y-B. Thyroid hormone-regulated early and late genes during amphibian metamorphosis. In: Gilbert LI, Tata JR, Atkinson BG, eds, Metamorphosis. Postembryonic Reprogramming of Gene Expression in Amphibian and Insect Cells. San Diego: Academic Press, 1996: 505-38.
81. Stolow MA, Shi Y-B. Xenopus sonic hedgehog as a potential morphogen during embryogenesis and thyroid hormone-dependent metamorphosis. Nucl Acids Res 1995; 23:2555-62.
82. Shi Y-B, Hayes WP. Thyroid hormone-dependent regulation of the intestinal fatty acid-binding protein gene during amphibian metamorphosis. Dev Biol 1994; 161:48-58.

83. Baker BS, Tata JR. Prolactin prevents the autoinduction of thyroid hormone receptor mRNAs during amphibian metamorphosis. Dev Biol 1992; 149:463-7.

84. Iwamuro S, Tata JR. Contrasting patterns of expression of thyroid hormone and retinoid X receptor genes during hormonal manipulation of *Xenopus* tadpole tail regression in culture. Mol Cell Endocrinol 1995; 113:235-43.

85. Ulisse S, Esslemont G, Baker BS et al. Dominant-negative mutant thyroid hormone receptors prevent transcription from *Xenopus* thyroid hormone receptor β gene promoter in response to thyroid hormone in *Xenopus* tadpoles in vivo. Proc Natl Acad Sci USA 1996; 93:1205-9.

86. de Luze A, Sachs L, Demeneix B. Thyroid hormone-dependent transcriptional regulation of exogenous genes transferred into *Xenopus* tadpole muscle in vivo. Proc Natl Acad Sci USA 1993; 90:7322-6.

87. Furlow JD, Berry DL, Wang Z et al. A set of novel tadpole specific genes expressed only in the epidermis are down-regulated by thyroid hormone during *Xenopus laevis* metamorphosis. Dev Biol 1997; 182:284-98.

88. Cruz-Reyes J, Tata JR. Cloning, characterization and expression of two *Xenopus* bcl-2-like cell-survival genes. Gene 1995; 158:171-9.

89. Boise LH, Gonzalez-Garcia M, Postema CE et al. bcl-x a bcl-2-related gene that functions as a dominant regulator of apoptotic cell death. Cell 1993; 74:597-608.

90. Yaoita Y, Nakajima K. Induction of apoptosis and CPP32 expression by thyroid hormone in a myoblastic cell line derived from tadpole tail. J Biol Chem 1997; 272:5122-7.

91. Sachs L, Abdallah B, Hassan A et al. Apoptosis in *Xenopus* tadpole tail muscles involves Bax-dependent pathways. Faseb J. 1997; in press.

92. Sachs L, Lebrun JJ, deLuze A et al. Tail regression, apoptosis and thyroid hormone regulation of myosin heavy chain isoforms in Xenopus tadpoles. Mol Cell Endocrinol 1997; 131:211-219.

Perspectives and Prospects

Two major reasons were given in the Foreword that justify this book: 1) the enormous current interest focused on early embryogenesis, and 2) that postembryonic development is characterized by its strict hormonal control. Analysis of postembryonic development is important for understanding the acquisition of the adult phenotype, while hormonal signaling mechanisms provide insight into how cell-cell and cell-environment interactions determine the late developmental changes. While it is impossible to even briefly cover all the diverse aspects of each of these two vast topics in a monograph of this size, it is hoped that the reader will have found it useful to consider them together in a single volume.

There are both differences and similarities in the phenomena—some apparent and some real—that set apart or unify the processes of postembryonic development and early embryogenesis. A major difference is the strong influence exerted by the external milieu of cells of multicellular organisms and of the environment of the organism on late stages of development. Various physical and chemical triggers that initiate postembryonic development are converted to hormonal signals. Hormones themselves are primitive substances and have been put to different uses during evolution. While nonprotein hormones have been highly conserved through evolution, it is significant that even peptide hormones and growth factors have also been structurally conserved. It is important to bear in mind these salient features (which have been the theme of the first two chapters) in designing and interpreting new experimental approaches to understanding postembryonic development.

Undoubtedly, receptors are the most important element of hormonal signaling mechanisms. The spectacular developments in the last 15 years of gene technologies have provided an enormous amount of information on two major classes of receptors, not only for hormones but for growth factors, vitamins, nutrients and xenobiotics.

Hormonal Signaling and Postembryonic Development,
by Jamshed R. Tata. © 1998 Springer-Verlag and R.G. Landes Company.

Both the membrane and nuclear receptors are derivatives of cellular homologs of oncogenes (as are many protein hormones and growth factors). Generally, but not exclusively, membrane receptors serve to initiate or coordinate rapid responses of a target cell to hormonal signals by the generation of intracellular second messengers. On the other hand, the cascade of responses originating from the interaction between nuclear receptors and their ligands is relatively slow and does not depend on the intervention of classical second messengers. Although a single hormone usually plays a predominant role, most postembryonic developmental processes, such as morphogenesis, tissue regression, oogenesis, and neural maturation, are regulated by multiple hormones. Hence the recent recognition of 'crosstalk' between hormones and receptors acquires increasing importance, particularly of that between one hormone acting through a membrane receptor and the other through a nuclear receptor. Clearly, the design of future studies on hormonal control of such processes as functional maturation of mammary gland, egg formation and programmed cell death will have to take into consideration the involvement of receptors in multi-hormonal controls. Another area of expanding future research stems from the applications of transgenesis and homologous recombination in extending our understanding of receptor function. If the recent findings of the phenotype (or its absence) exhibited by knock-out mice in which a hormone receptor gene is deleted are anything to go by, future investigations based on homologous recombination techniques are bound to provide us with many surprises. It is difficult to predict how new studies will change our present comprehension of hormone receptor function, but change they will. Here one will have to accept the idea, on the one hand, of redundancy among receptors, and that they may serve a non-ligand dependent function, on the other.

Nuclear receptors are intimately associated with the regulation of gene expression, a key mechanism underlying all the manifestations of postembryonic development. Hence chapter 4 was devoted to the hormonal control of transcription. Again, the applications of techniques of gene cloning, cell transfection, X-ray crystal and NMR analysis of DNA-protein interactions, to name a few, have had an enormous impact on our understanding of nuclear receptors as ligand-activated transcription factors. These have allowed investigators to pin-point the significance of such properties as homo- and hetero-dimerization of the steroid and non-steroid subgroups of nuclear receptors, the high degree of specificity of the sequences of response element in their interaction with DNA-binding domains of the receptor and the conformational changes induced by the ligand. It was only a matter of time until co-activators, co-repressors, integrators would be discovered, as revealed by recent studies. In the short-term, many more such complexes of nuclear receptors with auxillary factors will be described. In the long-term, studies on the formation of multi-component regulators of eukaryotic gene transcription will have to address the question of tissue specificity. The differential responses of different cell types to the same hormone acting through the same receptor is a salient feature of developmental hormones. An unexpected recent finding is the property of the unliganded thyroid hormone receptor acting as a general transcriptional repressor. Will more such examples be found? Although for several years some investigators have alluded to the importance of considering the role of nuclear receptors in the context of their organization into the higher order of chromatin structure, it is only within the last year or two that a molecular handle has been provided by enzymes known to add or remove acetyl residues in histones. The significance of these recent findings on histone acetylase and deacetylase is that it emphasizes the nucleosome and chromatin as dynamic components of the transcriptional machinery. Only future work will confirm whether or not these early studies will lead to a major shift in our thinking about how hormones control transcription.

A major drawback in studying the molecular mechanisms underlying postembryonic or fetal development in mammals is the complications introduced by hormonal and other signals of maternal and placental origin. Hence the popularity of metamorphosis in hormonally manipulating free-living insect and amphibian larvae to dissect the cascade of regulatory events determining all major later developmental processes, such as morphogenesis, gene switching and programmed cell death. The ease of initiating and arresting these developmental changes in tissue culture and at different developmental stages with the use of hormones has generated much valuable information about expression of 'early' genes and downstream target genes. One chapter has thus been devoted to metamorphosis. The power of genetics in studying metamorphosis in *Drosophila* has been most valuable, but the failure until now of applying techniques of homologous recombination has been a major impediment. How relevant is metamorphosis in vertebrates to mammalian fetal and postnatal development? A survey of literature clearly shows that several processes exhibit similarity, e.g., hemoglobin switching, central nervous system maturation, re-structuring of the digestive system, etc. However, all of these seem to be 'exaggerated' or 'amplified' during metamorphosis.

Among the most prominent 'early' or direct-response genes activated during hormonal induction of metamorphosis are the genes encoding the receptors of the inducers themselves. It should be emphasized that most studies on autoinduction of hormone receptors in insect and amphibian metamorphosis have been restricted to mRNAs. It is essential to confirm that there is a matching upregulation of functional receptor protein, as a few studies have indicated. If this requirement is fully met then it raises the questions of how widespread is the phenomenon of receptor autoinduction and its functional significance. The answer to both questions seems to be affirmative and sets hormone receptor expression in developmental systems apart from the downregulation commonly associated with adult tissues or hormones that do not regulate growth and development. As to the physiological significance of auto-upregulation, one hypothesis put forward is that it indicates a dual threshold phenomenon whereby the low level of constitutively expressed receptor has to be elevated in order to activate (or repress) downstream target genes. Only more direct evidence, such as that based on gene knockout and direct measurement of affinities of receptor, or receptor complexes, with other receptors and cofactors for promoters of target genes that specify a particular developmental phenotype, can support or refute this hypothesis.

Most postembryonic developmental processes are regulated by an interplay between multiple hormones; these include metamorphosis, mammary gland maturation and oogenesis. One chapter (the seventh) has been devoted to considering the possibility that cross-regulation of receptor gene expression may underlie the multi-hormonal interplay. This has been illustrated here with amphibian vitellogenesis and metamorphosis. An immediate requirement is to analyze more hormone-dependent late developmental systems to establish how widespread is the phenomenon of receptor cross-regulation. At the same time further work is necessary to characterize receptor gene promoters, and also to determine that cross-talk between membrane and nuclear receptor pathways involves the regulation of expression of the latter class of receptors.

Finally, no discussion of hormonal signaling and postembryonic development is complete without considering programmed cell death so that the last chapter of this book is devoted entirely to this topic, largely based on the extensive tissue regression during amphibian metamorphosis. It is now clear that developmental stage- and tissue-specific apoptosis is an integral and essential part of postembryonic development. Much interest has been aroused recently in the identification of cell death effector and cell survival genes and their high degree of evolutionary conservation. An intriguing idea is the equilibrium shifting towards a death or survival pathway determined by a direct mutually neutralizing interaction between the protein products of the death effector and survival genes. Among several interesting questions that come up in the context of hormonally regulated development are whether these genes are directly responsive to hormones and their receptors, and whether the heterodimeric complexes of death and survival gene products are indeed relevant to this type of development. It would also be important to resolve whether or not the same mechanisms underly tissue restructuring by selective apoptosis, as in the maturation of central nervous system, digit formation and intestinal re-organization, as those determining whole organ histolysis such as the loss of tadpole tail and gills during metamorphosis. Our current knowledge of the molecular and cellular biology of programmed cell death may represent only the tip of the iceberg. To a large extent our understanding of developmental cell death will depend on the progress made in identification of more cell death and survival factors and the interplay among these.

Each of the eight chapters has tried to convey a brief historical background, current thinking and a few outstanding problems to be resolved. It is hoped that bringing together the two areas of hormonal signaling and postembryonic development in the way it has been done with the author's bias and selectivity will offer the reader some new insight into endocrinology and developmental biology. The book will have served its function if it helps the researcher interested in hormone action and postembryonic development in designing or interpreting future investigations in this fascinating area of biology.

Index

A

α-fetoprotein, 8, 112, 116
Accessory sexual tissue, 7-8, 19, 115-116,
144, 180, 187-188
Acquisition of metamorphic compe-
tence, 134
Actinomycin D, 64, 66, 68-69, 192-193
Activation of Vit genes, 157-158, 160-161
Activation of zygotic genes, 3
Activin, 5
Adenyl cyclase, 34
Adenylate cyclase, 34-35
Adrenal, 17, 22, 66, 154-155
Adult intestine, 195
Adult male *Xenopus* liver, 73
Adult phenotype, 5, 8, 93, 101, 117, 179, 211
AGGTCA hexad, 46
Albumin, 8, 15, 72, 75, 102-103, 105, 107, 110,
112, 116, 129, 131, 136, 162
All-trans retinoic acid, 33
Ambystoma, 111-112, 133
Amphibian metamorphosis, 7, 9, 14-16,
64, 94, 96-98, 100, 105, 107, 109, 111-112,
114, 162, 164, 168, 170, 180, 184, 186,
188-189, 191, 193, 196, 205, 214-215
Androgen, 7, 16, 40, 42-43, 46, 63-64, 115,
139, 141, 143-144, 172-173, 186-189
Anti-androgen, 188
Anti-estrogen, 74, 188
Antimetamorphic action of PRL, 162
AP-1, 37, 39, 49, 77, 79, 166
Apoptosis, 103, 179-182, 184, 188, 196-197,
203-204, 215
Ashburner model, 145
Assembly of the *Xenopus* vitellogenin B1
gene, 83
Auto-upregulation, 124, 139-140, 143,
145-146, 153, 163, 200, 214
Autocrine, 5-6
Autoinduction, 124-129, 131, 133, 135-136,
138-139, 141, 143, 145-147, 156-158,
160-163, 165-166, 168, 170-171, 190, 196,
200, 214
Autoinduction of nuclear receptors, 124,
146, 171
Autoinduction of TRβ mRNA, 165, 168,
196
Autophosphorylation, 32

Axolotl, 111, 133

B

Basal metabolic rate, 19, 21, 64, 68, 154
Basal transcriptional activity, 83, 139
Basal transcriptional machinery, 82-83
Basement membrane, 194-196
Bayliss and Starling, 13
bcl-2, 181-182, 186, 188, 201
Behavioral imprinting, 113
Berthold's experiment, 186-187
Biochemical differentiation, 101
Bone development, 113
Brown, 3, 105, 128, 135-136, 197, 200

C

c-myc, 182, 201
Caenorhabditis elegans, 181
Carbomyl phosphate synthetase, 75, 102,
105
Caspase, 182
CBP, 77, 79
CCAT box, 75
cDNA cloning and sequencing, 40
Cell adhesion molecule, 195
Cell death effector mRNA, 203
Cell proliferation, 102-103, 108
Cell survival gene, 182-183, 200-201,
204-205, 215
Cell-cell interaction, 6-7, 194-195
Cellular differentiation, 3-4
Central nervous system, 6-7, 13, 96, 114,
144, 190, 214-215
Chironumus, 104
Chromatin condensation, 180, 182
Chromatin structure, 49, 63, 82-85, 213
Chromosomal puff, 64, 100, 102, 104, 145
cis-element, 78
9-cis retinoic acid, 33
Cloning, 31-32, 40, 42, 68, 71, 75, 155
Co-activator, 77, 80, 82-83, 166, 172, 213
Co-repressor, 80-83, 166, 172, 213
Collagenase, 103, 199
Corticotrope, 22, 154-155
Corticotropin, 21, 94, 96, 153, 155
Corticotropin releasing factor, 21, 94, 96
COUP-tf, 42, 80